Freeman Clarke Coffin

The graphical solution of hydraulic problems

Treating of the flow of water through pipes, in channels and sewers, over weirs, etc.

Freeman Clarke Coffin

The graphical solution of hydraulic problems
Treating of the flow of water through pipes, in channels and sewers, over weirs, etc.

ISBN/EAN: 9783744741774

Printed in Europe, USA, Canada, Australia, Japan

Cover: Foto ©berggeist007 / pixelio.de

More available books at **www.hansebooks.com**

THE GRAPHICAL SOLUTION OF HYDRAULIC PROBLEMS.

TREATING OF THE FLOW OF WATER THROUGH PIPES, IN CHANNELS AND SEWERS, OVER WEIRS, ETC.

BY

FREEMAN C. COFFIN,

Member of the American Society of Civil Engineers.

FIRST EDITION.

FIRST THOUSAND.

NEW YORK:
JOHN WILEY & SONS.
LONDON: CHAPMAN & HALL, LIMITED.
1897.

ROBERT DRUMMOND, ELECTROTYPER AND PRINTER, NEW YORK.

CONTENTS.

CHAPTER	PAGE
I. Introduction	1
II. Units and Symbols	4
III. Flow of Water in Pipes	6
Formulas	6
Small Pipes, $\frac{1}{2}$–3 Inch	8
Explanation of Diagrams	8
Examples	9
Large Pipes, 4–60 Inch	10
Explanations and Examples	10
Problems Outside of Range of Diagrams	11
Compound Mains	14
Complex Mains	16
Old or Tuberculated Pipe-lines	24
IV. Flow of Water through Short Tubes	31
Formula and Coefficients	31
Examples	32
V. Rectangular Weirs	36
Formulas	36
Explanation of Diagrams	37
Examples	38
Wide Crests	40
Velocity of Approach	41
VI. Flow of Water in Channels	44
Kutter's Formula	44
Values of n	45
Explanations and Examples	47

CHAPTER	PAGE
VII. PIPE-SEWERS	50
Formula and Coefficient	50
Explanation of Diagram	50
Examples	50
VIII. FIRE-STREAMS AND DISCHARGE OF HOSE-NOZZLES	52
Height of Streams and Head Required	52
Discharge of Hose-nozzles with Head or Pressure Indicated at the Hydrant	53
Discharge of Hose-nozzles with Head or Pressure Indicated at Base of Play-pipe	53
IX. MISCELLANEOUS PROBLEMS	55
Horse-power of Falling Water, also Horse-power Required to Raise Water	55
Capacity and Size of Pumps	56
Coal Required for Pumping Water	58

TABLES.

Table No. 1. Equivalents of U. S. Gallons	60
" " 2. Discharge in Gallons per Minute for Different Velocities in Circular Pipes	62
" " 3. 11/6 and 6/11 Powers of Numbers	65
" " 4. Coefficients of Friction in Old Pipes	66
" " 5. " " Discharge in Old Pipes	66
" " 6. " " Velocity of Approach for Weirs	67
" " 7. Areas and Values of r in Circular Channels	68
" " 8. " " " " r in Rectangular Channels	69
" " 9. With Side Slopes of 1 to 1	70
" " 10. " " " " 2 to 1	71
" " 11. " " " " 3 to 1	72
" " 12. Equivalents of Pounds Pressure and Feet Head	73
" " 13. Height and Discharge of Fire-streams	74

CONTENTS. v

DIAGRAMS.

Diagram No.   1. Flow of Water through Pipes ⅜–3 Inches in Diameter.
"   "   2–17. Flow of Water through Pipes 4–60 Inches in Diameter.
"   " 18, 19. Flow through Short Tubes or Entry Head.
"   "   20. Value of c in $V = c(rs)^{\frac{1}{2}}$.
"   "   21. Weirs with Wide Crests.
"   " 22–24.   "   " Thin   " .
"   " 25, 26. Flow of Water in Channels.
"   "   27.   "   in Pipe-sewers.
"   "   28. Horse-power of Water.
"   "   29. Size and Capacity of Pumps.
"   "   30. Coal Required in Pumping.
"   " 31, 32. Discharge of Hose-nozzles.
"   " A and B. Experiments on Old Pipes.

GRAPHICAL SOLUTION
.OF
HYDRAULIC PROBLEMS.

CHAPTER I.

INTRODUCTION.

It is not the purpose of this book to discuss the laws governing the flow of water, the hydraulic experiments that have been made, or the formulas derived from them. The object of the book, as conceived by the author, is to provide a convenient instrument or tool for the practising engineer (who is already familiar with hydraulic laws and formulas) with which he can solve quickly and correctly the commonly occurring hydraulic problems by means of diagrams and with a minimum of calculation, either mental or written.

The diagrams are constructed upon well-known formulas, using coefficients that are generally accepted as safe for the conditions to which they are intended to apply. While it was intended to have the computations correctly made and carefully checked, no attempt has been made to secure mathematical refinements, such being in most cases entirely out of place in hydraulic problems. The computations for these diagrams were made with the slide-rule and logarithmic cross-section paper.* Such computations are correct to two figures, and sometimes three or four, and for these problems this is a greater degree of accuracy than can usually be secured in all of the data.

In regard to the possibility of obtaining sufficiently accurate results from the diagrams themselves, this can be said: That in all problems relating to the flow of water through short tubes, long pipes, sewers, and channels, the result can be read from the diagram with a greater degree of precision than the engineer is warranted in relying upon in practice. The variations in the commercial sizes of pipes will probably cause greater errors than those that result from inability to read the diagram closely. Differences that cannot be estimated in the surfaces of pipes, sewers, or channels will produce differences in result beside which any errors from the above sources will be insignificant. Details of actual construction, small variations from the design, differences in the character of surfaces and many other elements, tend to render a refinement in figures absurd.

Of the subjects relating to the flow of water treated in this book, it is possible that in weirs alone a few cases will arise where the diagrams will not give results sufficiently close; but such cases are only those in which the weir has been constructed in perfect conformity with the conditions from which the formulas were derived.†

---

* The paper used was that designed by Mr. John R. Freeman. M. Am. Soc. C. E., with four complete squares of 10-inch base on one sheet.

† Weirs are excepted from the general statement because the conditions of their construction are more under the control of the engineer than those of the other classes.

There are, however, probably few weirs used for ordinary purposes that are so constructed in every detail. If the diagrams are based upon proper formulas, the computations correctly made, and the drawing carefully done, results can be read from them with sufficient precision for all practical purposes.

The diagrams are offered to the profession in the hope that they may prove useful, as similar ones have been a valuable aid to the author for several years. Any of the usually occurring hydraulic problems can be solved by them, either in field or office, with no other tables, formulas, or information than those contained in this book, and in nearly every case without the use of pencil or paper.

CHAPTER II.

UNITS AND SYMBOLS.

In choosing the units to be used, preference is given to those in most general use.

Measures of quantity of water are always given in U. S. gallons per minute. Whenever possible, a second scale gives cubic feet per second.

Table No. 1, page 60, gives equivalents of U. S. gallons per minute.

All linear dimensions are given in feet and decimals, except in the case of diameters of pipes, which are given in inches.

SYMBOLS.

The following characters will always have the meaning here assigned, unless otherwise stated :

$a =$ area in square feet ;
$c = \begin{cases} \text{coefficient in weir formula,} \\ \text{coefficient in Chezy formula ;} \end{cases}$
$D =$ diameter in feet ;
$D' =$ depth of water in channels in feet ;
$d =$ diameter in inches ;
$g =$ acceleration of gravity ;
$h =$ head in feed for pipes and weirs ;
$l =$ length in feet of pipes and weirs ;

UNITS AND SYMBOLS.

$n =$ Kutter's coefficient for character of surface of pipes and channels;

$p =$ wetted perimeter in feet;

$Q =$ cubic feet of water per second;

$q =$ U. S. gallons of water per minute;

$R =$ radius $= \dfrac{D}{2}$;

$r =$ hydraulic mean radius of channels and pipes $= \dfrac{a}{p}$;

for circular pipes running full $r = \dfrac{D}{4}$;

$s =$ hydraulic slope for pipes and channels $= \dfrac{h}{l}$;

$v =$ mean velocity in feet per second;

for circular pipes running full $v = \dfrac{Q}{a}$;

$W =$ width of channels.

Effective head and loss of head by friction in long pipes is given on diagrams as "friction-head in feet."

CHAPTER III.

FLOW OF WATER IN PIPES.

Problems connected with the flow of water in pipes are of constant occurrence in the practice of the hydraulic engineer.

They may be divided into three general classes:

(I) The size (D), length (l), and head (h) being given to find the quantity (Q) or discharge.

(II) The size, length, and quantity being given to find the required head or loss by friction.

(III) The quantity, head, and length of pipe line being given to find the size of pipe required.

There are many formulas, based upon numerous experiments, in use for the solution of these problems. Without discussing the respective merits of these, it may be said that the Chezy formula is used as the basis of computation for the diagrams relating to this subject. This formula was selected on account of its simplicity, and because it seemed possible with a varying coefficient to meet the conditions of different diameters of the pipes and of varying velocities more nearly than by any other form. This formula is expressed as follows:

$$\text{I.} \quad v = c(rs)^{\frac{1}{2}};$$

$$\text{II.} \quad s = \frac{v^2}{c^2 r};$$

$$\text{III.} \quad r = \frac{v^2}{c^2 s}$$

With these three forms of the equation, the value of either factor of the problem can be found when the other two are given.

Note.—When r is known, the diameter of a circular pipe running full can be obtained by multiplying r by 4.

The values of the coefficient c used in these diagrams are those given by Hamilton Smith, Jr., in his work on Hydraulics. These values seem to have been derived with great care from the most reliable hydraulic experiments. The reader is referred to the above-named book for further information on this subject.* These values of c are given graphically on diagram No. 30. The full curves are copied from Smith's Hydraulics; the dotted curves are interpolated by the author.

Although the Chezy formula is very simple, there is this practical difficulty in using it in the calculation of individual problems, in the second form in which it is given, i.e., $v = c(rs)^{\frac{1}{2}}$: that c cannot be taken correctly until v is known, and an approximation must first be made by taking a value of c according to some assumed or estimated value of v. One or two approximations will generally bring the result. This objection to the formula does not hold, however, in constructing diagrams, as the first form, or $s = \dfrac{v^2}{c^2 r}$, or, reduced to the simpler expression, $s = \dfrac{1}{r}\left(\dfrac{v}{c}\right)^2$ is the one best adapted for such computations.

ACCURACY OF FORMULAS FOR THE FLOW OF WATER IN PIPES.

It should be borne in mind that, while it is desirable that formulas should give with accuracy values for cer-

---

* "Hydraulics," by Hamilton Smith, Jr. (John Wiley & Sons).

8 GRAPHICAL SOLUTION OF HYDRAULIC PROBLEMS.

tain given conditions of pipes, exactly these conditions rarely exist in practice. Therefore too much weight must not be given to mathematical refinements in ordinary practice, however desirable these refinements are in experimentation. When, for instance, the value of the coefficient c is in the vicinity of 100, a variation in the coefficient in several formulas of 4 or 5 should not be considered as invalidating either for practical purposes, as perhaps the difference in the commercial sizes of pipes of the same nominal diameter will cause a greater variation. Differences in the interior surfaces of the pipes, that with our present knowledge are impossible of exact estimation, will produce a greater variation than that caused by such differences of the coefficient. Therefore, while it is desirable to use the best formula for clean, straight pipes as a basis, a sufficient margin should be allowed for these differences of condition. The subject of the effect of the age of pipe lines is discussed elsewhere in this chapter.

Explanation of Diagrams.

SMALL PIPES.

The diagrams for the flow of water in small pipes (those from three eighths of an inch to three inches in diameter) are constructed for a length of 100 feet. Results for other lengths can be taken proportionately.

These sizes are plotted upon one sheet, diagram No. 1.

The diameter of the pipes is given by curves originating at the upper left-hand corner, the discharge in gallons per minute by a vertical scale on the right of sheet, the velocity by dotted curves, and the loss of head from friction by a horizontal scale at the top of the sheet.

Examples of Use.

CASE I. Size, length, and discharge given to find head.

Example: What is the loss of head in a 1-inch pipe 75 ft. long, delivering 16 gallons per minute?

On diagram No. 1, at the intersection of the line representing 16 gallons of water with the curve representing 1-inch pipe, find the friction in feet, viz., 22 ft.; this is in a length of 100 ft.; therefore $22 \times \frac{75}{100} = 16.5$ ft., the required answer.

CASE II. Size, length, and head given to find discharge.

Example: What is the capacity of a 2-inch pipe 350 ft. long with effective head of 40 feet?

To use this diagram the length must be reduced to 100 ft., and the head in like ratio; thus $40 \times \frac{100}{350} = 11.4$ per 100 feet in length.

At intersection of 11.4 friction-head and 2-inch curve find capacity in gallons = 70 gallons per minute, flowing with velocity of 7 ft. per second.

CASE III. Length, head, and discharge given to find size.

Example: What is the size of pipe necessary to discharge 175 gallons per minute through 500 ft. in length with a head of 80 ft.?

$80 \times \frac{100}{500} = 16$ ft. per 100 ft. in length. The intersection of 175 gallons and 16 feet head is between curves of $2\frac{1}{2}$ and 3 inches diameter, and of the usual sizes, the latter must be used.

LARGE PIPES.

For pipes larger than 3 inches in diameter there is a diagram for each size. The size is given in the upper right-hand corner of diagram-sheet. On these diagrams the length of pipe is given by a vertical scale on the right-hand side; the friction-head by a horizontal scale on top and bottom of sheet; the quantity in gallons per minute by oblique lines radiating from a common origin at the upper left-hand corner. Cubic feet per second are not given on these diagrams, but may be taken from the Table No. 1 of equivalents of U. S. gallons per minute on page 60. The velocity in feet per second is given by a heavy broken line parallel with the oblique line of quantity. These diagrams are more convenient than those for small pipes, because the solution of all problems within a wide range can be read from them directly for the proper length without multiplication or division.

Examples of Use.

Case I. Size, length, and effective head given to find discharge.

Example: What is the discharge of a 14-inch conduit-line 7500 ft. long with an effective head of 30 ft.?

Ans. 1860 gallons per minute; velocity 3.9 ft. per second. (From diagram No. 7 for 14-inch pipe.)

Case II. Size and length of pipe and discharge given to find loss of head or friction.

Example: What is the loss of head in 5500 ft. of 12-inch pipe delivering 1100 gallons per minute?

The answer will be found at intersection of lines representing 1100 gallons and 5500 ft. of pipe on diagram No. 6 for 12-inch pipe. It is 18.2 ft. head. The velocity is 3.1 ft. per second.

CASE III. Length, head, and required discharge given to find size of pipe.

Example: What is the size of pipe necessary to discharge 2500 gallons per minute through 7500 ft. of pipe with 30 ft. head?

The diagrams show that a 14-inch pipe will discharge 1860 gallons, and a 16-inch pipe will discharge 2675 gallons; and of the usual commercial sizes, the latter is the one that must be used.

These diagrams cover in their range nearly all of the problems met in practice. If the solution of problems that do not come within their limits is desired, it can be found indirectly as follows.

Problems Outside of Limits of Diagrams.

When the length of line is greater than that given on the diagram.

CASE IV. Size, length, and discharge given to find the friction-head.

Divide the given length by some number that will give a quotient within limits of the diagram; find head for this length, and multiply by the number used to divide length.

Example: What is the friction-head in 12,000 ft. of 12-inch pumping main, with a delivery of 1000 gallons per minute, or velocity $= 2.85$?

$12,000 \div 2 = 6000$; friction in 6000 ft. $= 16.7$ ft.; $16.7 \times 2 = 33.4$ feet head.

CASE V. Size, length, and head given to find discharge.

Divide both length and head by the same number, and find discharge for the quotient, which will be the answer.

12 GRAPHICAL SOLUTION OF HYDRAULIC PROBLEMS.

Example: What is the discharge through 15,000 ft. of 10-inch pipe with head of 50 ft.?

$15,000 \div 2 = 7500$ ft.; $50 \div 2 = 25$ ft. head; discharge from diagram for 25 ft. head through 7500 ft. of pipe $= 680$ gallons; $v = 2.8$.

When the length given is less than that for which it is convenient to read results on the diagram.

CASE VI. Size, length, and discharge given to find head.

Multiply length by any convenient number, and divide the resultant head by same number.

Example: What is the loss of head in 36-inch penstock 400 ft. long, discharging 25 cu. ft. per second or 11,200 gallons per minute?

$400 \times 100 = 40,000$ ft.; 11,200 gallons through 40,000 ft. of 36-inch pipe $= 39.2$ ft. head; $39.2 \div 100 = 0.392$ head.

CASE VII. Size, length, and head given to find discharge.

Multiply both head and length by any convenient number, and find discharge corresponding to the products.

Example: What is the discharge of a 30-inch pipe 200 ft. long with an effective head of 0.4 ft.?

$200 \times 100 = 20,000$ ft. long; $0.4 \times 100 = 40$ ft. head; the discharge of 30-inch pipe through 20,000 ft. with 40 ft. head $= 10,000$ gallons; vel. $= 4.5$.

Note.—The entry and velocity heads in the last two examples, an important element, should be considered in all cases of comparatively short lines. This subject is treated in the next chapter.

Where the discharge or head is greater than that given on diagram.

CASE VIII. Size, length, and discharge given to find head.

Divide the given discharge by any number that will bring it within the limits of the diagram. Multiply the friction due to the quotient by the 11/6 power of the number used.*

A table of 11/6 and 6/11 powers is given on page 65.

Example: What is the loss of head in 500 ft. of 6-inch pipe discharging 1500 gallons per minute?

1500 ÷ 2 = 750; 750 gallons through 500 ft. of 6-inch pipe = 24 ft. head; 24 × ($2^{11/6}$ = 3.56) = 85.5 ft., answer.

CASE IX. Size, length, and head given to find discharge.

Divide the given head by the 11/6 power* of any number that will bring head within limits of diagram. On given length find discharge due to this quotient; multiply this discharge by the number taken.

---

* Although the exponent of the formula $v = c(rs)^{\frac{1}{2}}$ is of the second power, the variation in the coefficient c renders the results practically identical with those of a formula of the 11/6th power in which c is constant for all velocities.

The reader is referred to a paper in the Journal of Associated Engineering Societies for June, 1894, by Mr. W. E. Foss, in which he shows that a formula of the 11/6th power is in practical accordance with the results of experiments. He proposes such a formula for the flow of water in pipes and channels which is very simple, and, with the tables given in the paper, renders the computation of such problems much less tedious than by most formulas in use.

See page 22 for remarks on the use of the second power instead of the 11/6th.

Example: What is the discharge through 2000 ft. of 8-inch pipe with effective head of 125 ft.?

$125 \div (2^{11/6} = 3.56) = 35.1$; discharge due to 35.1 ft. head $= 925$ gallons $\times 2 = 1850$ gallons per minute.

Where the discharge or the head is less than can be easily read from the diagram.

CASE X. Size, length, and discharge given to find head.

Multiply the given discharge by a number that will give a product large enough to be easily read from the diagram; divide the head obtained by the 11/6 power of that number.

Example: What is the head in a 16-inch pipe 5000 ft. long, delivering 100 gallons per minute?

$100 \times 10 = 1000$; head for 1000 gallons $= 3.4$ ft; $3.4 \div 10^{11/6} = 0.05$ ft.

CASE XI. Size, length, and head given to find discharge.

Multiply the given head by the 11/6 power of some number; divide the discharge due to the product by the number.

Example: What is the discharge of a 16-inch pipe 6500 ft. long with a head of .25 ft.?

$0.25 \times 10^{11/6} = 17$; discharge due to head of 17 ft. $= 2120$; $2120 \div 10 = 212$ gallons per minute.

COMPOUND MAINS.

A compound main is a single line of pipe which is not of uniform diameter for its entire length. It is composed of two or more sizes in one line.

The friction in such a main can easily be computed when the discharge is given.

FLOW OF WATER IN PIPES.

To find the discharge when the total head is given is more difficult.

The finding of both the friction and discharge is much facilitated by the use of these diagrams.

CASE I. Size, length, and discharge given to find head.

Find head for each size separately, and add results for total head.

Example: What is the head in a compound main composed of 3000 ft. of 12-inch pipe and 7000 ft. of 16-inch, delivering 1500 gallons per minute?

Head in 3000 ft. of 12-inch pipe = 17.4 ft.
" " 7000 " " 16- " " = 9.6 "

Total head, 27.0 " *Ans.*

CASE II. Size, length, and head given to find discharge.

Assume some quantity for the discharge; find the head due to such discharge in the given line, the same as in Case I. Divide the correct head by this head, and multiply the assumed discharge by the 6/11 power of the quotient.

Example 1: What is the discharge of a compound main of 4500 ft. of 10-inch pipe and 7000 ft. of 16-inch pipe with a total effective head of 40 ft.?

Assume a discharge of 1000 gallons.

Head in 4500 ft. of 10-inch pipe = 30.2 ft.
" " 7000 " " 16 " " = 4.6 "

Total, 34.8 "

Then $1000 \times \left(\dfrac{40}{34.8}\right)^{6/11} = 1000 \times 1.075 = 1075$ gallons, correct discharge.

Check on above calculation:

1075 gals. in 4500 ft. of 10-inch pipe = 34.7 ft. head.
"   "   " 7000 "   " 16- "   "   = 5.3 "   "
                            Total,   40.0 "   "

Example 2: What is the discharge of a compound main of 10,000 ft. of 16-inch, 6000 ft. of 12-inch, and 2000 ft. of 10-inch pipe with total head of 50 ft.?

Assume discharge of 1500 gallons per minute.

Head in 10,000 ft. of 16-inch = 13.8
 "   "   6,000 "   " 12   " = 34.9   [VIII, page 13.
 "   "   2,000 "   " 10   " = 28.1, by method of Case

Total head for assumed discharge, 76.8

Then $1500 \times \left(\dfrac{50}{76.8}\right)^{6/11} = 1184$ gallons, correct discharge.

COMPLEX MAINS.

The problems involved in the flow of water in complex mains, or a system of mains in which the water flows to the point of discharge through two or more lines of pipe of different sizes and lengths, are of great importance in practice and should receive careful attention. Owing to the immense amount of work required to solve them by the ordinary methods, they are undoubtedly left unsolved in many cases.

These problems arise in nearly all cases where an efficient fire system is to be designed for a town or city, or where the efficiency of one already constructed is to be ascertained. A valuable and suggestive paper on this subject was read by Mr. John R. Freeman, M. Am. Soc. C. E., before the New England Water-works Association, and published in the Journal of that Association, Vol. VII.

The diagrams for the flow of water in long pipes provide a ready means of solving such problems, and the following examples are given of their use.

CASE I. System of piping and discharge given to find head.

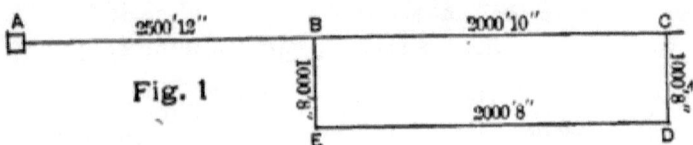

Fig. 1

Example: With reservoir at *A* and arrangement of piping as shown in Fig. 1, what will be the loss of head with a draft of 1000 gallons per minute at *C*?

First, find head in single line $AB = 7$ ft.

The flow will divide at *B*, part of it going through *BC* and the remainder through *BEDC*.

It is self-evident that the total friction or loss of head must be the same in one line as in the other, or the flow from the two lines would meet at *C* under different pressures, which is impossible.

Assume a loss of head for these lines, and on the diagrams find the discharge due to this assumed head.

Thus assume 20 ft. head; then

discharge in *BC*, 2000 ft. of 10 in. = 1240 gals.
" " *BEDC*, 4000 " " 8 " = 470 "

Total discharge for both lines, 1710 "

Divide the given discharge by the discharge of the assumed head, and multiply the 11/6 power of the quotient by the assumed head, and the product will be the true head.

$$\left(\frac{1000}{1710}\right)^{11/6} \times 20 = 0.37 \times 20 = 7.4, \text{ true head.}$$

Add head in AB $\qquad\qquad = 7$

Total loss of head, 14.4 for 1000 gals.

(*a*) If, instead of assuming 20 ft. head, 8 ft. head had been assumed, then

discharge in $\quad BC = 755$ gals.
" $\qquad$ " $BEDC = 280$ "

Total assumed discharge, 1035 "

$$\left(\frac{1000}{1035}\right)^{11/6} \times 8 \ = 7.50 \text{ feet.}$$

Add head in $AB = 7 \qquad$ "

Total loss of head, $\quad 14.50$ "

Note.—The slight variation in the results of the two examples are due to slight errors in reading diagrams.

(*b*) After the correct head for one rate of discharge is obtained, it is very easy to find it for any other by the following formula:

$$\left(\frac{\text{required discharge}}{\text{known discharge}}\right)^{11/6} \times \text{known head} = \left\{\begin{array}{l}\text{head for re-}\\\text{quired dis-}\\\text{charge.}\end{array}\right.$$

Thus when draft of 1000 gallons through a given system of pipes causes a loss of head of 14.50 ft. as above, other drafts would be as follows:

$$\left(\frac{500}{1000}\right)^{11/6} \times 14.50 = \ 4.06 \text{ ft.}$$

$$\left(\frac{1200}{1000}\right)^{11/6} \times 14.50 = 20.3 \quad \text{"}$$

$$\left(\frac{1500}{1000}\right)^{11/6} \times 14.50 = 30.6 \quad \text{"}$$

CASE II. System of piping and head given to find discharge.

Example: With same arrangement of piping as in Case I, what will be the discharge with total head of 30 ft.?

First assume the head in the lines BC and $BEDC$; thus assume 15 ft.

BC = 2000 ft. of 10 in., discharge = 1060 gals.
$BEDC$ = 4000 " " 8 ",   "   = 400 "

Total discharge with 15 ft. head,   1460 "

The loss of head in line AB, or 2500 ft. of 12 in.,
due to a discharge of 1460 gallons = 13.9 ft.
Add head in BC and $BEDC$ = 15 "

Total assumed head, 28.9 "

To find the discharge due to the given head use following formula:

$$\left(\frac{\text{given head}}{\text{assumed head}}\right)^{6/11} \times \text{assumed discharge} = \text{true discharge.}$$

$$\left(\frac{30}{28.9}\right)^{6/11} \times 1460 = 1.02 \times 1460 = 1488 \text{ gallons,}$$

the true discharge for 30 ft. head.

The most complex problems can be solved by the application of the foregoing principles.

Where there are a number of mains, with cross-mains between, it will of course require the exercise of thought and judgment to decide which lines are effective and how they should be classified. The experienced engineer can usually make a very close approximation in such cases without going into the detail of every line, and thus avoid much tedious work.

20 GRAPHICAL SOLUTION OF HYDRAULIC PROBLEMS.

The following case is given as an illustration of a possible treatment of a rather intricate problem.

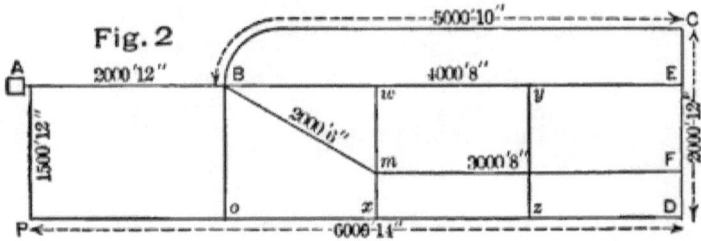

Fig. 2

With the above arrangement of piping and reservoirs at A, what will be the loss of head caused by a draft of ten 1-inch* hydrant streams or 2000 gallons per minute from hydrants on line CD?

As the hydrants will be fed both ways through short lines, the friction in line CD can safely be ignored, although if the location of the hydrants were fixed it could be computed.

The cross-lines Bo, wx, yz can also be ignored. The slight effect that they would have would be on the safe side.

Assume the loss of head in lines BC, BE, and BmF; it will be the same in the three.† One of these lines, BmF, is a compound main, and the discharge will be assumed here to start with. Assume 200 gallons; then

in line Bm, 2000′ of 6″, head = 8.6 feet
" " mF, 3000′ " 8″, "   = 3.2 "
                              Total, 11.8 "

---

* See Table No. 12 for definition of fire-steam.

† This would only be true in case the pressure in the line CEF was the same at all points; it is assumed here that it is practically so.

Therefore the discharge with assumed head of 11.8 feet will be:

in line BC, 5000' of 10", = 562 gals.
" " BE, 4000' " 8", = 353 "
" " BmF = 200 "

Total discharge through BC,
BE, and BmF,   1115 "

Loss of head in AB due to flow of 1115 gallons:

1115 gallons through 2000' of 12", = 6.8 ft.
Add for BC, BE, and BmF,   11.8 "

Total head for assumed discharge
in system $ABCF$,   18.6 "

It is evident that the loss of head in line APD must be equal to that in $ABCF$. As APD is a compound pipe, assume flow to be 1000 gallons. The head due to this flow is as follows:

In AP, 1500' of 12", 4.2 head
" PD, 6000' " 14", 7.8 "

Total, 12 "

Then discharge due to 18.6 feet, or head in $ABCF$, is

$$\left(\frac{18.6}{12}\right)^{6/11} \times 1000 = 1270 \text{ gallons per minute.}$$

Therefore total discharge with loss of head of 18.6 ft. is:

In system $ABCF$, 1115 gallons per minute
" " APD, 1270 "   "   "

Total, 2385

Finally, to find loss of head due to the required discharge of 2000 gallons per minute, use the formula

$$\left(\frac{\text{required discharge}}{\text{assumed discharge}}\right)^{11/6} \times \text{assumed loss of head};$$

or

$$\left(\frac{2000}{2385}\right)^{11/6} \times 18.6 = .728 \times 18.6 = 13.50 \text{ ft. head, } ans.$$

If it were desired to find loss of head due to a draft of ten 1¼-inch fire-streams or 2500 gallons per minute, or any other quantity, use the same formula.

If the discharge for a given loss of head is required in the above system, use the formula

$$\left(\frac{\text{given head}}{\text{assumed head}}\right)^{6/11} \times \text{assumed discharge} = \begin{cases} \text{required} \\ \text{discharge.} \end{cases}$$

In the above case, suppose that a head of but 15 feet were available, or that it were not desirable to allow of a greater loss of head; then

$$\left(\frac{15}{18.6}\right)^{6/11} \times 2385 = .86 \times 2385 = 2050 \text{ gallons,}$$

the discharge for a loss of head of 15 ft.

Note.—In all of the cases given in this chapter the 2d power instead of the 11/6, and the 1/2 power instead of the 6/11, could be used; and if the assumed quantities closely approximated the real ones, the difference in results would be very small. The following table shows the difference in a few cases.

When the friction caused by a draft of 1000 gals. is 14.50:

Gals.	Head by 11/6 power.	Head by 2d power.
1000	14.50	14.50
500	4.06	3.62
1200	20.30	21.00
1500	30.60	32.60
2000	51.60	58.00

When a head of 10 feet would cause a discharge of 1000 gallons, the difference for different heads is shown by the following table:

Head.	Discharge by 6/11 power.	Discharge by 1/2 power.
10	1000	1000
5	685	707
12	1100	1095
15	1245	1225
20	1458	1415

The advantage of using the 2d power instead of the 11/6 is that the formula can be computed on the slide-rule with one setting, which would be as follows:

For formula

$$\left(\frac{\text{required discharge}}{\text{assumed discharge}}\right)^2 \times \text{assumed head} = \text{true head}$$

set assumed discharge on lower slide over required discharge on lower scale, and over assumed head on upper slide read true head on upper scale.

For formula

$$\left(\frac{\text{given head}}{\text{assumed head}}\right)^{1/2} \times \text{assumed discharge} = \text{true discharge}$$

set assumed head on upper slide under given head on upper scale, and under assumed discharge on lower slide find the true discharge on lower scale.

OLD OR TUBERCULATED PIPE-LINES.

It is now generally recognized that the loss of head caused by the friction in pipes increases with the age of the pipes. There is, however, no generally accepted formula that takes into account the factor of age.* Kutter gives a variable coefficient n for roughness of the sides of channels. In old pipe-lines there are generally no data relating to the condition of these surfaces, and the age of the line is the only easily determined element. While this is not so satisfactory as the known condition of the pipes in question would be, which is affected by the character of the water and of the coating as well as by the age of the line, there are no other data relating to this condition given in the recorded experiments upon old pipes. It is desirable to have some ready means of approximately estimating the effect of age upon pipe-lines, as the formulas given for clean pipe cannot be safely used for old lines without correction. Although good judgment should always be used, some mathematical assistance is needed.

COEFFICIENTS OF LOSS OF HEAD.

Wishing to provide some systematic means of correcting the results given by the diagrams for clean pipes, so that they may be safely used for old lines, the author made a study of all the experiments upon old pipes that were available to him. As might be expected, the results of these are conflicting and do not furnish a satisfactory basis for a mathematical formula.

---

* Since the above was written a valuable book by Edmund B. Weston, "The Friction of Water in Pipes" (D. Van Nostrand Co.), giving the friction caused by the flow of water in pipes, has been published, which gives coefficients for friction in old lines.

The graphical method seemed to be the most satisfactory one for arriving at some conclusion. Accordingly, the results of the experiments expressed as a percentage of the excess of loss of head by friction over the loss under the same condition in new pipes, as computed by the formula adopted in this book, was plotted as shown on diagram A, using the horizontal scale for the percentage of excess of friction, and the vertical scale for the velocity of flow in feet per second. The result of each experiment of a series is represented by a small circle, and all results of a series are connected by a broken line. By graphical comparison, curves * that seemed to the author to represent the average results of the experiments were drawn for different ages of pipe. It may be seen that with a few exceptions the general direction of the broken lines is inclined from the lower left- to the upper right-hand corner. This indicates that the percentage of increase of friction for any age of pipe is not constant, but increases with the increase in velocity.

The rate of this increase (or the inclination of the broken lines) differs greatly in the experiments. The full lines drawn to represent this are inclined in the ratio of one horizontal to two vertical, and represent an increase in the percentage which varies with the square root of the increase in velocity.

This rate almost coincides with that of one of the most carefully made experiments, viz., that upon a 48-inch pipe by Mr. Desmond Fitz-Gerald. It is also an approximate average of all the experiments.

It is indicated by the experimental results that the excess of friction is greater in small pipes than in large

---

* As the base of the diagram is logarithmic, these curves are straight lines.

ones, other conditions being the same. There are not sufficient data to make any classification based upon size.

In giving the value to the different ages, an endeavor was made to use the approximate average for all sizes. The curve for fifteen years nearly coincides with the above-named experiment on a 48-inch pipe eighteen years old.

Table No. 4, giving coefficients by which to multiply the friction loss in new, clean pipe, was compiled from diagram A by adding the percentage of excess of friction-loss, as shown by the curves for several ages of pipe, to unity.

For example, the excess of head in a pipe ten years old with a velocity of flow of 2 ft. per second is 30 per cent; adding this to 100 per cent (the friction-loss in clean pipe) makes 130 per cent, or written as a coefficient, 1.30.

COEFFICIENTS OF DISCHARGE.

Diagram B was designed as a means of deriving coefficients of velocity or discharge from diagram A. This is also constructed upon a logarithmic base. The horizontal scale represents the loss of head by friction; the vertical scale, velocity in feet per second. Line CD represents the friction in new, clean pipe when a velocity of one foot causes a loss of head of one foot (it is immaterial what value is given to the loss of head caused by a velocity of one foot) as calculated by the formula used in this book. Lines $C'D'$, $C''D''$, etc., represent the friction in pipes of different ages. The small circles through which these lines are drawn were obtained by multiplying the loss of head given at the intersection of CD with line of velocity upon which the circles are plotted, by the proper coefficient from diagram A. For example, on velocity = 2 the loss of head given at CD is 3.56 ft.; this × (14 per cent from diagram A at $v = 2$ and age = 5 years + unity = coef. 1.14) = 4.06 ft., the loss of head in pipes five years old with $v = 2$.

The coefficients of discharge are derived as follows: On vertical lines EF drawn through the intersection of velocity lines with CD, find the velocity at the intersection of $C'D'$, $C''D''$, etc., as for example: On vertical EF drawn through intersection of $v = 3$ with CD find the velocity at $C''D'' = 2.55$. This is the velocity in a pipe ten years old under the conditions and with the head that produces a velocity of 3 ft. in new pipes. Then $\frac{2.55}{3} = 0.85 =$ coefficient of discharge in a pipe ten years old. The coefficients in Table No. 5 were derived in this manner. The velocities in the first column are those that would be produced in new, clean pipes under the conditions of the problem, as given by the diagrams for flow of water in clean pipe.

It is not claimed that there is mathematical exactness in these coefficients as applied to any particular case. It is hoped that they represent an approximate average of the results of recorded experiments, and may, when intelligently used, be of some assistance in the solution of problems relating to old lines.

Following is a list of the experiments the results of which were used, with references to the sources from which they are taken.

Experimenter.	Place.	Pipe. Diameter.	Pipe. Age, Years.	Source from which Data were taken.
Fitz-Gerald	Boston.....	48″	18	Trans. Am. Soc. C. E., vol. xxxv.
Darcy......	Paris.......	10″	old	
Ehmann...	Stuttgart..	10″	6	
Iben.......	Hamburg..	12″	12	Ganguillet & Kutter,
"	" ..	12″	2	Flow of Water, translated
"	" ..	12″	14	by Hering and Trautwine.
"	" ..	12″	15	
"	" ..	12″	22	
"	" ..	16″	25	
Leslie.....	Edinburgh.	15″	30	Hamilton Smith's Hydraulics.
"	" ..	16″	8 or 9	
Weston ..	Providence	6″	4	
Simpson ..	" ..	12″	7	
" ...	" ..	12″	Unknown	
" ...	" ..	12″	4	Flow of Water in Pipes.
" ...	" ..	19″	13	Weston, Trans. Am. Soc. C. E., vol. xxii.
" ...	" ..	30″	3	
Green......	Brooklyn...	36″	Heavily tuber-culated	
Gale.......	Glasgow...	48″	8	
Forbes.....	Brookline..	14″ and 16″	18	Journal N. E. Water-works Assn., vol. vi.
"	" ..	16″	18	
Hastings...	Cambridge.	30″ and 36″	8	Same, vol. viii.
Coffin......	Randolph..	14″	8	See following table.

Note.—All experiments were upon cast-iron pipe.

RESULTS OF EXPERIMENTS UPON A PIPE-LINE EIGHT YEARS OLD AT RANDOLPH, MASS.
2,700 ft. 14″ Pipe.

No.	Vel.	Loss of Head by Friction.			
		Actual.	Computed.	Per Cent of Computed.	Per Cent in Excess.
1	.80	1.50	2.85	53	
2	1.40	6.95	7.90	88	
3	1.82	13.80	12.70	108	8
4	2.09	18.30	16.40	111	11
5	2.34	23.50	20.30	116	16
6	2.50	28.00	22.70	123	23
7	2.80	34.00	28.	121	21

Examples of Use of Tables Nos. 4 and 5.

CASE I. When the discharge is given and the head required in an old line.

Example : What is the friction-head in a 20-inch pipe line 10,000 ft. long, 15 years old, with a discharge of 2000 gallons per minute?

Find velocity and friction-head for clean pipe from diagram No. 10: $V = 2$; friction-head $= 7.6$. On Table No. 4 find opposite $V = 2$ and under 15 years the coefficient $= 1.47$; then $7.6 \times 1.47 = 11.2$ feet-head.

(*a*) What is the head in the above example with a discharge of 5500 gallons per minute?

Friction-head for clean pipes from diagram $10 = 49.5$; $V = 5.6$ ft., coefficient from Table No. $4 = 1.80$; $49.5 \times 1.80 = 89$ ft.-head.

CASE II. When the head is given and the discharge required.

Example : What is the discharge of a 10-inch pipe-line one mile long, 20 years old, with head of 40 feet?

From diagram No. 5 find discharge $= 1070$ gallons; $v = 4.4$. On Table No. 5, opposite $V = 4$ to 5, and under 20 years, find coefficient 0.72; then $1070 \times 0.72 = 770$ gallons.

For compound and complex mains use the same methods, taking an approximate average of the velocities in the different sizes.

As an approximate method and for preliminary work the coefficients for an age of 20 years and velocity of 3 feet per second may be memorized and used. They are 1.80 for loss of head and 0.75 for discharge.

An age of 20 years and velocity of 3 feet are very close approximations to the values used in ordinary practice.

Note.—Nothing has been said in this chapter about the loss of head caused by bends, branches, and gates in pipe-lines. The author is of the opinion that in lines of new pipes with the ordinary velocities the friction loss as given on the diagrams will include all loss of head in clean-coated pipe-lines as usually laid.* The coefficients for old lines are intended to apply to lines of pipes with the usual special castings and gates.

In designing all engineering works it is customary to use a factor of safety. There should be no exception to this custom in the design of pipe lines or systems. It cannot be considered good practice to design a line that will according to the formula used discharge exactly the required amount of water. For instance, if it is *absolutely necessary* that a line shall discharge ten million gallons of water per day, safety would require that a diameter be chosen which, when computing the discharge as carefully as possible, will deliver at least eleven millions per day, or an increase of 10 per cent. This is a small factor of safety as compared with that used in other branches of engineering.

It is generally the case that the amount supposed to be required is only an approximate estimate of that actually required. In such, as in all cases, the engineer must use his judgment gained from study and experience. No formula, table, or diagram, however correct within their limitations, can alone assure successful design. They are but useful tools for the skilful workman.

Note.—For sizes of pipes not given on diagram, take area and value of r from table No. 7 and find solution on diagram No. 25 for channels, using $n = .011$.

---

* See paper on Friction in Several Pumping-mains in the Journal of the N. E. W. W. Assoc., vol. x, No. 4.

CHAPTER IV.

FLOW OF WATER THROUGH SHORT TUBES.

The flow of water through short tubes is treated in this book on account of its relation to the entry-head of long pipes. By a short tube is understood a pipe that has a length equal to about three times its diameter. The loss of head caused by the flow of water through short tubes includes the loss due to friction at entrance and the head required to generate velocity, and will be designated simply as entry-head hereafter in this book. This loss is constant in any given pipe for any given velocity without regard to the length of the pipe-line. Consequently it must be determined separately from the loss of head caused by friction in a long pipe which is in direct proportion to its length.

In very long pipe-lines the entry-head is an insignificant element in the problem, but in comparatively short lines, with high velocities, it is an important factor, and should not be ignored.

The diagrams for entry-head are constructed from the formula

$$v = o(2gh)^{\frac{1}{2}}, \quad \text{or} \quad h = \frac{V^2}{2go^2},$$

with the following value for o as given in Hamilton Smith's Hydraulics:

32 GRAPHICAL SOLUTION OF HYDRAULIC PROBLEMS.

Circular pipes:

Mouth flush with side of reservoir.......... 0.825
" projecting into reservoir, with square ends........................... 0.715
Bell-shaped mouthpiece, small velocities..... 0.950
" " " large velocities...... 0.995

$2g$ was taken as 64.36.

There are two diagrams on this subject:

No. 1 for pipe from 4 to 20 inches in diameter;
No. 2 " " " 20 " 60 " " "

The diameter of the pipe is shown by a curve from the upper left-hand corner; the loss of head in feet by horizontal scales; the one at the top of the diagram for pipes with the mouth flush with the side of the reservoir; that at the bottom for pipes projecting into the reservoir and with square ends. These are the types most in use. If it is required to find the loss for bell-shaped mouthpieces, multiply the head for flush pipes, or that given at the top of the diagram, by 0.75. The scale of gallons per minute is given on the right-hand side of the diagram, and the scale of cubic feet per second on the left.

Examples.

CASE I. Diameter of pipe and discharge given, to find the head.

Example : What is the loss of head in a 10-inch pipe discharging 4500 gallons per minute or 10 cu. ft. per second when pipe is flush

Ans. (from diagram No. 18, top scale), 7.7 ft.

(a) What would be the loss of head in the above example, if the pipe projected into reservoir?

Ans. (from lower scale), 10.25 ft.

(b) If the mouth of the pipe were bell-shaped?

Ans. 7.7 ft. loss for flush pipe × 0.75 = 5.77 ft.

CASE II. Diameter and head given, to find discharge in a short tube or pipe.

Example: What will be the discharge through a short tube 12 inches in diameter, flush ends, under head of 4.2 ft.?*

Ans. (using upper scale of diagram No. 18), 4850 gallons per minute.

If the end projects into reservoir, in above example, take the head, 4.2 ft. on lower scale, and the discharge will be 4190 gallons per minute.

CASE III. Available head and required discharge given, to find diameter of pipe.

Example: What size of pipe is required to discharge 2000 gallons per minute with flush ends and head of 2 ft.?

Using upper scale of diagram No. 18 for head, the intersection of $h = 2$ and $q = 2000$ falls between the curves of 8 and 10 inch pipes; therefore of the usual commercial sizes a 10-inch pipe would be required.

CASE IV. The diameter of a long pipe-line and its discharge given, to find total loss of head.

Find the loss due to friction in the pipe by the methods of Chapter III, and add to that amount the loss of head caused by entry and velocity found on the diagrams for entry-head, as in Case I: the sum will be the total head.

Example: What is the total loss of head in a pipe-line

---

* Head is measured from surface of water to centre of pipe.

14 inches in diameter, 1000 ft. long, flush end, discharging 4000 gallons per minute?

The friction loss, taken from diagram No. 7 for flow of water in long pipes, will be. 16.35 ft.
The loss for entry, as shown on diagram No. 18 for entry-head, will be.......... 1.60 "
                                                Total, 17.95 "

CASE V. The diameter and length of a long pipe and the available head given, to find discharge.

This problem cannot be solved directly, as the loss of head caused respectively by the entry and the friction is not known.

The methods described in Chapter III, Compound Mains, Case II, may be adopted, and a discharge assumed.

Example: What will be the discharge of a 16-inch pipe 1000 ft. long, with an effective head of 6 ft., with flush end?

Assume a discharge of 3000 gallons per minute; for this quantity, by diagrams

No. 8, the friction-head will be......... 4.95 ft.
No. 18, the entry-head will be.......... 0.5 "
                                    Total, 5.45 "

Then, following method of Chapter III, Case II, of Compound Main,

$$\left\{ \frac{\text{given head}}{\text{head due to assumed discharge}} \right\}^{1/2} \times \text{assumed discharge}$$
$$= \text{true discharge, or } \left(\frac{6}{5.45}\right)^{1/2} \times 3000 = 3150^* \text{ gallons.}$$

---
* See note on page 22.

FLOW OF WATER THROUGH SHORT TUBES.   35

The loss of head due to each cause can be taken from the diagram after true discharge is found. In above example, for a discharge of 3150 gallons,

from diagram No. 8, friction-head = 5.40 ft.
"   "   "   " entry-head   = .58 "
                                        ─────
                                Total,   5.98 " *

Note.—It will be found advantageous in such cases as the above, where the pipe is comparatively short, to find the friction for a longer length; in this case the friction-head can be found more closely by finding it for 5000 ft. instead of 1000, and dividing result by 5.

The diagrams for entry-head are very convenient for finding loss of head in circular penstocks and other short pipe-lines. In such cases both the friction and the entry-head must be considered.

---

* The entry-head increases as the second power of the increase of velocity, and when it is as great as the friction-head the second power will be as correct as the 11/6.

CHAPTER V.

RECTANGULAR WEIRS.

The diagrams for the discharge of weirs are computed from the well-known Francis' formula, $Q = clh^{3/2}$. For heads above 0.5 ft., $c = 3.33$.

For heads below .5, $c =$ the values given in the following table compiled from the paper of Fteley and Stearns,[*] which describes their experiments on low heads:

VALUE OF c FOR LOW HEADS IN $Q = clh^{3/2}$.

Head.	c.	Head.	c.
0.5'	3.33	0.2'	3.388
0.4	3.337	0.15	3.430
0.3	3.353	0.1	3.528
0.25	3.368	0.06	3.750

END CONTRACTIONS.

For each end contraction deduct $.1h$ from the length, or $.2h$ where there is contraction at both ends.

Note.—Hamilton Smith, Jr., gives a tables of values for the coefficient c in the formula $Q = \frac{2}{3}c(2g)^{\frac{1}{2}}lh$.[†]

---

[*] Published in vol. XII, Transactions of the Am. Soc. Civil Engineers.

[†] This formula can be reduced to the form of Francis' formula by substituting values of c and $2g$.

RECTANGULAR WEIRS.

Values are given for weirs both with and without end contraction.

These coefficients give results slightly different from those obtained by Francis' formula (from one to three per cent), and it is probable that in cases where the weirs are constructed entirely in accordance with the given conditions that the results may be more accurate than with a constant coefficient, even where a correction is made for end contraction, as above.

See "Hydraulics," by Hamilton Smith, Jr., for full discussion of the subject and description of experiments.

Smith says that no weir measurements of water should be made with h less than 0.2' where accuracy is essential.

EXPLANATION OF WEIR DIAGRAMS.

Very little explanation is necessary for an understanding of these diagrams. They are three in number:

No. 1 gives lengths from 0 to 4 ft.
" heads   " 0.01 " .5 "
    discharge " 0 " 800 gals. per min.
"   "   "   0 " 1.77 cu. ft. per sec.

No. 2 gives lengths from 0 to 8 ft.
" heads   " 0.05 " 1.50 ft.
" discharge " 0 " 8000 gals. per min.
"   "   "   0 " 17.7 cu. ft. per sec.

No. 3 gives lengths from 0 to 16 ft.
" heads   " 0.1 " 2.5 " *
" discharge " 0 " 40,000 gals. per min.
"   "   "   0 " 89 cu. ft. per sec.

---

* Caution: It is not safe to rely implicitly upon the results with heads greater than 2 feet. Mr. Francis considered the limits of his formula to be from 0.5 to 2 feet.

The lengths of the weirs are given in feet by a vertical scale on the side of the diagrams; the horizontal scale at the top gives discharge in gallons per minute; the scale at the bottom in cubic feet per second, and the oblique lines radiating from the upper left-hand corner give the head in feet.

Examples.

Example 1: What is the discharge of a weir 3 ft. long, with head of 0.26 ft., with contraction at each end?

First find corrected length for end contraction, or $3 - .2h = 2.948$ ft. On diagram No. 22, at the intersection of lines representing the above values, viz., $l = 2.948$ and $h = 0.26$, find discharge $= 597$ gallons per minute, or 1.325 cu. ft. per second.

Example 2: In the above example, if the head is 0.45, what will be the discharge?

Corrected length $= 3 - (.2h = .09) = 2.91$ ft.

It will be seen that the intersection of $l = 2.91$ and $h = 0.45$ does not fall on diagram No. 22; therefore look for it on No. 23. It is 1315 gallons per min., or 2.9 cu. ft. per second.

Diagram No. 23 is on a smaller scale than No. 22, and if it is desired to determine the discharge more closely by using No. 22, proceed as follows: Divide the corrected length by 2 or any convenient number; thus, $2.91 \div 2 = 1.455$; find the discharge for this length $= 655$ gallons; multiply this result by 2, or the number used to divide length, $655 \times 2 = 1310$.

On one or other of the three diagrams results for any head up to 1.40 ft. and any length up to 16 ft. can be read directly; also any head up to 2 ft. with lengths under 9.5 ft.

If it is required to find results where the intersection of l and h does not fall on the diagram that it is desired to use, or if the length is greater than that given on any of the diagrams, advantage can be taken of the principle that the discharge is proportional to the length, and the result found as in Example No. 2 (*a*).

N. B.—If there is end contraction, the length must be corrected for it before dividing.

Example 3: What is the flow over a weir 20 feet long with contraction at each end and head of 0.2 ft?

Corrected length, $20 - (.2h = .04) = 19.96$ ft.

To find result on diagram No. 22, where the length is but 4 feet: $19.96 \div 5 = 3.992$ ft. = discharge with $h = .2$ of 554 gallons per min. Multiply this by the divisor used, $554 \times 5 = 2770$ gallons per min., total discharge.

It is often desirable to know how long to construct a weir in order to discharge a certain maximum quantity of water and not exceed a certain head.

Example 4: What should be the length of a weir to discharge 2000 gallons per min., h not to exceed 0.6? On diagram No. 23 at intersection of $h = .6$ and $q = 2000$, find length $= 2.96$. A weir 3 feet long will meet the requirements, including correction for end contraction.

All problems connected with rectangular weirs can be solved by these diagrams, and with sufficient accuracy for all except the most careful work.

CONDITIONS TO WHICH THE DIAGRAMS APPLY.

The conditions of weirs to which the formula and diagrams apply are as follows:

The crest must be horizontal, the sides vertical, and the plane of the face of the weir must be approximately at right angles to the line of flow. The edges of the weir must be so thin that the escaping water shall only come in contact with the up-stream corner edges. For accurate measurement stiff metallic plates with smooth, straight edges should be used. For weirs where there is no end contraction the sides of the canal should be prolonged down-stream beyond the crest, but should not extend below its level, as this would prevent the access of air under the escaping vein. For full contraction the water must discharge freely into the air.

The air must be given free entrance under the escaping vein.

The head should be measured 6 feet up-stream from the weir upon a stake or fixed point set level with its crest. For weirs with end contractions the sides of the weir openings should not be less than twice the maximum head from the sides of the canal, and the same distance from the bottom. The length of the weir should not be less than three times the maximum head. When the sheet falls on an apron, the latter should be not less than the total head below the crest. If the discharge falls into a body of water the latter may be level with the crest.* The error will not exceed one per cent from this cause if the weir is submerged to the extent of 15 per cent of the head on the crest.†

WIDE CRESTS.

It often happens that measurements are made over weirs with wide crests.

Messrs. Fteley and Stearns made a series of experiments on the flow over wide crests, which are described

---

* Smith says that if air is admitted freely behind the falling sheet the surface of water may be nearly to level of crest.

† See "Submerged Weirs" in the paper of Fteley and Stearns referred to above.

in the paper already referred to. As the result of their work they give a table of corrections to be applied to the depth or head on a wide crest in order to obtain the depth on a sharp crest which will pass the same quantity of water. The results of their table as worked out are given in the form of a percentage of the discharge over a thin-plate weir on diagram No. 21. To use this diagram take the discharge due to the measured head from the diagrams for thin crests in the usual manner, and multiply it by the percentage given on diagram No. 21 for given width of crest and head.

Example: What is the discharge of a weir 5 ft. long, without end contraction, and with head of 0.55 over a crest 6 inches wide?

The discharge from the diagram No. 23 for this crest = 3050 gallons per minute. Multiply this by the percentage from diagram No. 21 for 6-inch crest with head of 0.55 ft. or $3050 \times 0.90 = 2745$ gallons per minute.

The diagram gives crests from 2 to 24 inches wide, and also gives the proper percentage for overflows of the type of the Lawrence dam worked out from Mr. James B. Francis' formula for that type, $Q = 3.01208 l h^{1.3}$.

This dam is 3 ft. wide on top, with a slope on the upstream side of about 3 horizontal to 1 vertical.

VELOCITY OF APPROACH.

When there is an appreciable current towards the weir the velocity of approach must be considered if accuracy is required.

This subject is thoroughly discussed in Smith's Hydraulics. In the paper of Fteley and Stearns, before referred to, is given a description of a series of experiments and their conclusions from them. Any one wishing to make extremely careful measurements is referred to the above, and also to Francis' work on hydraulics, "The Lowell Experiments."

For ordinarily careful work Table No. 6 gives corrections to add to the measured head before taking results from the diagram. This table is based upon the coefficients given by Fteley and Stearns for weirs with and without end contractions, and for different heads and depths below the crest.

The mean velocity of approach is found by dividing the approximate discharge over weir by the area of cross-section of the channel. The head due to this velocity is found by the formula $h = \dfrac{v^2}{2g}$ when $v =$ mean velocity of approach and $2g = 64.36$.

The coefficients of Fteley and Stearns vary from 1.33, to 1.87 for weirs without, and 1.88 to 2.42 for those with, end contraction, and are used to multiply h as found above, and the product is added to the measured head.

The method of using the table is as follows: Find the discharge in cubic feet per second due to the measured head from the diagram; divide this discharge by the area of the cross-section of the channel, to find mean velocity of approach;* find the nearest number to velocity as found in column 1 of Table No. 6, and opposite it, under the column giving the conditions nearest those of the case, find the correction in feet to be added to the measured head; finally, find on the diagram the result due to the corrected head.

Example: What will be the discharge over a weir 4 ft. long with end contraction and measured head of 0.6 ft.?

---

* The velocity may also be obtained by observations with floats or a current-meter.

The channel is 6 ft. wide and 1.5 ft. deep = 9 sq. ft. area, depth below crest 1 foot.

Corrected length = $4-(.2h$ or $.12) = 3.88$ ft.; the approximate discharge, from diagram No. 23, due to .6 head is 6 cu. ft. per second. Velocity of approach $= 6 \div 9 = 0.667$ feet per second. On Table No. 6, for weirs with end contraction, opposite $v = 0.7$, and in column for depth below crest $= 1$ foot and head of 0.60, find the correction $= 0.013$; add this to the measured head, $= 0.6 + 0.013 = 0.613$.

Then, with corrected length $= 4 - 2h = 3.877$, the discharge for corrected head of $0.613 = 6.2$ cu. ft. per second or 2780 gallons per minute.

Note.—Fteley and Stearns give a diagram in the paper referred to from which the head to be added for velocity of approach can be taken directly, without first finding the mean velocity. This diagram only applies to weirs without end contraction.

CHAPTER VI.

FLOW OF WATER IN CHANNELS.

The engineer meets many problems connected with the flow of water in channels of other than circular sections. There is such variety in the shape of these sections that it is impracticable to construct diagrams for any special forms, ranging as they do from an egg-shaped sewer to a canal of trapezoidal section in earth. Instead of trying to treat these forms (except that of circular pipe-sewers) on special diagrams, a general diagram is constructed which gives the values of $v =$ velocity per second, $s =$ slope or fall per 1000, and $r =$ hydraulic mean radius. When any two of these three elements of the problem are known, the third can be read from the diagram. Thus all problems of flow in channels of any section can be readily solved.

The basis of these diagrams (Nos. 25 and 26) is the Chezy formula: $v = c(rs)^{1/2}$, as in the case of circular pipes. For channels, however, the value of the coefficient c is found by Kutter's formula, which is

$$\left\{ \frac{\dfrac{1.811}{n} + 41.6 + \dfrac{.00281}{s}}{1 + \left(41.6 + \dfrac{.00281}{s}\right)\left(\dfrac{n}{\sqrt{r}}\right)} \right\},$$

all symbols having the meaning given in Chapter II.

This formula seems better adapted to channels than any other, on account of the different values that Kutter gives to n corresponding with different degrees of roughness of the surface of the channel. These values are given in the following table:

VALUES OF n IN KUTTER'S FORMULA.

Character of Surface.	Value of n.
Well-planed timber	0.009
Plastered with neat cement } Also glazed pipe }	.010
Plastered with mortar composed of one part sand to three of cement	.011
Unplaned timber and uncoated C–I pipe	.012
Ashlar and first-class brickwork	.013
Second-class brickwork and well-dressed stone	.015
Rubble in cement in good order	.017
" " inferior condition, and very firm and regular sides in gravel	.020
Rivers and canals in good order, free from vegetation	.025
Same, having stones and weeds occasionally	.030
" in bad condition, with much vegetation	.035

There are two diagrams for channels: No. 25, for those having surfaces corresponding to Kutter's $n =$ .011 and .013; No. 26, for surfaces corresponding to $n =$ .025 and .030. On these diagrams the slope or fall per 1000 is given by curves from the upper left hand; the hydraulic mean radius, or r, by a vertical scale on each side; and the velocity, or v, by a horizontal scale. On diagram No. 25 the upper scale is for $n =$ .011 and the lower scale for $n =$ .013; on diagram No. 26 the upper scale is for $n = .025$ and the lower one for $n = .030$.

For other values of n the following table gives the percentage to use. These percentages are only a mean of those for different slopes and values of r; they are usually not more than two or three per cent in error. For $n = .015$ and .017 and .035 they are about 10 per cent in error for values of $r = .1$ and 4; while for $r = 1$ they are about 3 per cent in error for all slopes and values of n. While it may not be considered that the table of percentages gives a very close result in some cases, it must be remembered that this is a subject that contains a very uncertain factor in the value of n. It is probable that the accuracy as given by the percentages is sufficient for all practical purposes of design or computation. When greater accuracy is desired, the reader is referred to the table giving values of coefficient c in $v = c(rs)^{1/2}$ in Hering and Trautwine's translation of "The Flow of Water," by Ganguillet and Kutter.* This gives the results of Ganguillet and Kutter's formula for said coefficient, worked out for different slopes and values of r and s.

TABLE GIVING PERCENTAGE OF v FOR DIFFERENT VALUES OF n.

For Diagram No. 25, upper scale

When $n = .009$, multiply or divide v by 1.25
" $n = .010$, " " " v " 1.10
" $n = .011$, " " " v " 1.00
" $n = .012$, " " " v " .90
" $n = .013$, " " " v " .80
" $n = .015$, " " " v " .70
" $n = .017$, " " " v " .60

* John Wiley & Sons.

FLOW OF WATER IN CHANNELS.

For diagram No. 26, upper scale.

When $n = .020$, multiply or divide v by 1.28
" $n = .025$, " " " v " 1.00
" $n = .030$, " " " v " .80
" $n = .035$, " " " v " .68

For other values of n than those given on diagrams:

When v and s are given, multiply v as found on the diagram by the percentage in the table. When r and v are given, divide v by the percentage before reading result from the diagram.

N. B.—Always use upper scale of diagram for values of v when using percentages.

Explanation and Examples.

In using the diagrams, the hydraulic mean radius of the channel, or r, must be calculated. When two of the factors r, v, and s are known, the third can be found by inspection of the diagrams. s is given as the fall in 1000.

CASE I. When r and s are given, to find v.

Example 1: What will be the velocity in a rectangular channel 6 ft. wide, 2 ft. depth of water, fall of 0.25 in 1000, with surface corresponding to $n = .011$?

Find the value of r:

$$r = \frac{a}{p} = \frac{6 \times 2}{6 + 2 + 2} = 1.2 \begin{cases} \text{or find area and value} \\ \text{of } r \text{ on Table No. 8.} \end{cases}$$

On diagram No. 25, at the intersection of $r = 1.2$ and $s = 0.25$ per 1000, find the value of v on upper scale $= 2.50$.

If the discharge is required, $Q = va = 2.50 \times 12 = 30$ cu. ft. per second.

(*a*) If, in the above example, the surface corresponds with $n = .013$, the velocity will be found on the lower scale, $= 2$.

(*b*) If the surface corresponds with $n = .015$, multiply v, as found in Example 1, by the percentage taken from table on page 46, or $v = 2.5 \times .70 = 1.75$.

Example 2: What will be the velocity in a canal of trapezoidal cross-section, 10 ft. wide on the bottom, side slopes 3 to 1, depth of water 5 ft., $s = 0.125$ per 1000, with surface corresponding to $n = .030$?

Find area and value of r for this section on Table No. 11. Area $= 125$; $r = 3.01$.

At intersection of $r = 3.01$ and $s = 0.125$, find velocity on lower scale, diagram No. 26, $= 1.12$ per sec.

CASE II. When r and v are given, to find s.

Example: What is the slope required in a rectangular channel 6 ft. wide, with depth of water of 1 ft., surface $n = .013$, to produce a velocity of 3 ft. per second?

On Table No. 8 find area $= 6$ ft., $r = 0.75$.

On diagram No. 25, at the intersection of $r = 0.75$ and $v = 3$, measured by lower scale, find the value of $s = 1.02$ per 1000.

If in the above example the surface corresponds with $n = .020$, what will be the required slope?

First divide v by the percentage from the table on page 47, for $n = .020 = \dfrac{3}{1.28} = 2.34$.

At the intersection of $r = 0.75$ and $v = 2.34$, on upper scale of diagram No. 26, find value of $s = 2.80$.

CASE III. When the quantity of water to be delivered and the slope is given, to find the size and shape of the channel.

The problem is met in this form in designing channels.

As v is involved in the quantity and area of the channel, this form of the problem cannot be directly solved by the diagram. Tables Nos. 7, 8, 9, 10, and 11 give the width and depth of rectangular and trapezoidal channels and diameters of circular ones, with areas and values of r, and will, with the diagrams 25 and 26, facilitate the solution of the above problem.

Example: What will be the width, depth, and velocity of a rectangular channel, with slope of 0.5 in 1000 and surfaces corresponding to $n = .013$, to deliver 30 cu. ft. per second?

It is evident that there are many combinations of width and depth that will meet the conditions. A preference may be had, however, for a certain approximate width. Assume that to be 6 ft.; there will also probably be a limit to the allowable velocity, say 3 ft.

This, with $Q = 30$, will set the minimum limit of area at 10 ft. Therefore $\frac{10}{W = 6} = 1.66$ deep; now, with $W = 6$ and $D = 1.66$, $r =$ approx. 1.1 from Table No. 8. With $r = 1.1$ and $s = 0.5$, $v = 2.68$, but with the above area v must $= 3$. D must be increased; try $D = 1.80$; then $r =$ approx. 1.13, area $= 10.8$, v from diagram $= 2.70$; $2.70 \times 10.8 = 29.2$ cu. ft. per second. If a close design is required, a trifling change in D will give the correct result.

In this way, by the use of the tables and diagrams, the suitable size and shape of a channel is easily found.

CHAPTER VII.

PIPE-SEWERS.

The discharge and velocity of pipe-sewers are given on diagram No. 27 for sizes from 6 to 24 inches, when running full; when running half full, velocity is the same, and the discharge is one-half. This diagram was constructed from Kutter's formula, and gives results for two values of the coefficient n. The scale at the top of the sheet gives the fall or slope in feet per 100 when $n = .011$. The scale at the bottom gives slope when $n = .013$. The diameter of the sewer is given by curves originating at the upper left-hand corner. The discharge in gallons per minute is given by a vertical scale on the right, and cubic feet per second by a vertical scale on the left. The velocity in feet per second is given by the broken curved lines.

EXPLANATION OF DIAGRAM.

Case I. Size of pipe and slope per 100 ft. given, to find velocity and discharge.

Example: What will be the discharge of a 10-inch pipe-sewer, running full, with a slope of 0.48 per 100, with surface corresponding to Kutter's $n = .011$?

At the intersection of diameter = 10 inch and slope = 0.48 per 100 taken on the upper scale, find the discharge = 790 gallons per minute, or 1.78 cu. ft. per second, with velocity = 3.25.

If in the above example the surface corresponds with $n = .013$, use the lower scale for slope, and the answer would be 650 gallons per minute; velocity = 2.60.

CASE II. The slope per 100 and required velocity or discharge being given, to find the size of the pipe.

Example: What size of pipe is required to secure a velocity of 3 ft. per minute when the slope is 0.32 per 100 and the surface corresponds with $n = .011$?

At the intersection of velocity = 3, and slope on the upper scale = 0.32, find next larger size of pipe = 12 inch.

If the required discharge is given instead of the velocity, the method of solution is the same.

CASE III. Size of sewer and required velocity or discharge given, to find fall per 100.

Example: What is the necessary fall in a 15-inch sewer to produce a velocity of 2.5 ft. per second = discharge 1375 gallons per minute, with surface corresponding to $n = .013$?

At intersection of $v = 2.5$ and diameter = 15 inch, find value of fall read from the lower scale = 0.24.

Note.—When it is required to find velocity more closely than can be read from velocity-curves, it may be done by reading the discharge instead and taking velocity from Table No. 2, which gives the number of gallons for different velocities in the different diameters.

Note.—For sizes of sewers, either pipe or brick, not given on diagram No. 27 use table No. 7 to find area and value of p, and solve problem by diagram No. 25 or 26, using such value of n as the case requires.

CHAPTER VIII.

FIRE-STREAMS AND DISCHARGE OF NOZZLES.

In designing pipe systems for fire-protection, it is desirable to know the pressure or head required at the hydrant to throw a stream to a given height, and the quantity of water required for such stream.

Mr. John R. Freeman, M. Am. Soc. C. E., in a valuable paper published in the Transactions of the Am. Soc. of Civil Engineers, vol. XXI, describes and gives the results of a great number of careful experiments made by him upon the pressure and discharge of fire-streams.* In this paper are tables giving the height and discharge of fires-treams under various conditions of pressure, length of hose, size and shape of nozzle, etc. Table No. 13 in this book is compiled from Mr. Freeman's tables and gives the head required at the hydrant to deliver fire-streams through lengths of hose varying from 50 to 500 ft.; the height to which the streams reach, and the number of gallons per minute required for each. The table gives the results for 1-, $1\frac{1}{8}$-, and $1\frac{1}{4}$-inch smooth nozzles with the ordinary best quality of $2\frac{1}{2}$-inch rubber-lined hose.

---

* A similar paper by the same author with the same tables was also published in the Journal of the N. E. Water-works Association, March 1890.

This table is intended for ready reference, and comprises those sizes of nozzles in most common use. The reader is referred to the paper mentioned above for more extended tables and a discussion of the subject.

Discharge of Hose-nozzles.—Diagram No. 31 is constructed from tables given in the paper referred to. It gives the discharge of hose-nozzles through 50 and 100 ft. of 2½-inch ordinary best quality of rubber-lined hose, and the head or pressure indicated by a gauge at the hydrant.

The diagram may be used when the approximate quantity of water delivered by a pump or flowing through a pipe is to be found. Mr. Freeman says that when the hose corresponds with the conditions of his tables the results are close enough for practical purposes, and when careful judgment is used to interpolate between the values of rough and smooth hose, results within 5 per cent can be obtained. The diagram only gives results for ordinary best quality rubber-lined hose with the inside smooth. When testing pumps or lines it is advisable to use that class of hose. Fifty feet of hose should be used in preference to one hundred. Full curves on diagram represent the size of the nozzles with 50 ft. of hose; the dotted curves with 100 ft.

Diagram No. 32 gives the discharge of nozzles and the head or pressure as indicated by a gauge at the base of the play-pipe. In this case the effect of friction in the hose is eliminated, and with smooth, true, carefully-calibred nozzles very accurate results can be obtained.*

This diagram is constructed from the formula $v = o(2gh)^{1/2}$, using $o = 0.99$.

---

* See paper on The Nozzle as an Accurate Water-meter, by J. R. Freeman. Trans. Am. Soc. Civil Engineers, vol. XXIV, 1891.

Mr. Freeman found this coefficient to be 0.995 as the result of his experiments.

On the diagram one curve represents two sizes of nozzles, the larger just double the diameter of the smaller, as $\frac{3}{4}''$ and $1\frac{1}{2}''$.

The scale of gallons is given for the smaller sizes; it must be multiplied by four for the larger ones.

CHAPTER IX.

MISCELLANEOUS DIAGRAMS.

UNDER this head there are several diagrams that the author has found useful in his practice, and that may be called hydraulic diagrams.

HORSE-POWER.

Diagram No. 28.—This diagram gives the horse-power for falling water, and also the horse-power required to raise water. The vertical scale at the right of the diagram gives the quantity of water in cubic feet per second; that at the left gives the same in gallons per minute. The horizontal scale at the top gives the horse-power as represented by the total energy or 100% efficiency; that at the bottom gives the same with an efficiency of 75%, or the amount usually taken as the efficiency of a good water-wheel.

CASE I. When the quantity of water and effective head are given to find the power.

Example: What is the total horse-power of 8.5 cu. ft. per second with an effective head on wheel of 13 ft.? At intersection of cu. ft. per sec. = 8.5 and head = 13, find value of horse-power as read on upper scale = 12.6.

(*a*) In the above example what would be the effective power from a wheel with an efficiency of 75%? The answer, read from the lower scale, = 9.4 horse-power.

CASE II. Power required to pump water.

When the quantity to be pumped and the total head to be pumped against are given to find the effective horse-power required.

Example: What is the power required to pump 2000 gallons per minute against a total head on the pump of 150 ft.?

Answer taken from upper scale = 76 horse-power.

DISCHARGE OF SINGLE PUMPS

Diagram No. 29.

This diagram is designed to give the capacity of reciprocating pumps or pumping-engines, either single, duplex, or triplex. The horizontal scale at the top and bottom of the sheet gives the capacity in gallons per minute; the horizontal scale between the two parts of the diagram gives the revolutions per minute. The vertical scale on the sides gives the piston-speed in feet per minute. The oblique lines radiating from the upper left-hand corner give the diameters of pistons and plungers in inches. Those from the upper right-hand corner give the length of stroke in inches.

The upper part of the diagram is for the smaller sizes of pumps, and the lower part for the larger ones. The diagrams are designed for single pumps with double-acting pistons or plungers. For duplex double-acting pumps multiply the discharge, as taken from the diagram, by 2. For triplex single-acting pumps multiply by $1\frac{1}{2}$.

CASE I. Given the diameter of plungers and piston-speed to find the discharge.

Example 1: What is the capacity of a single-acting pump with a piston 10″ in diameter and a piston-speed of 80 feet per minute, diameter of piston-rod 2 inches?

At the intersection of lines representing the above values of speed and diameter find the discharge as follows:

Discharge of 10″ plunger, 80′ piston-speed = 326 gallons
Less 1/2 discharge of 2″ rod at same
  piston-speed, from diagram, = $\frac{14}{2}$ = 7   "
                                              ─────
                                      Total, 319   "

(*a*) If the above were a duplex pump, multiply by 2: 319 × 2 = 638 gallons per minute.

When the length of stroke and revolutions per minute are given instead of piston-speed, first find, at the intersection of line representing the number of revolutions with that of length of stroke, the line representing piston-speed, and follow this line to its intersection with that of the given diameter, and read the discharge at that point as before.

Example 2: What is the discharge of a duplex pump with 12-inch plungers and 18-inch stroke at 35 revolutions per minute, diameter of piston-rod 3 inches?

At the intersection of 35 revolutions and 18-inch stroke find piston-speed (107 ft.); follow this to intersection of 12-inch diameter and there read gallons

                                        = 620 gals. per min.
Less one half rod same speed = $\frac{40}{2}$ = 20   "   "   "
                                            ─────
                                  Total, 600 "   "   "

As it is a duplex pump, multiply by 2 = 1200 gallons.

CASE II. When the required discharge and piston-speed are known, to find the diameter of plunger.

Example 1: What size of plunger is necessary in a duplex pump to discharge 4000 gallons per minute; the piston-speed not to exceed 130 ft., diameter of rod $3\frac{1}{2}$ inches?

As it is a duplex pump, first divide 4000 by 2 = 2000 + loss caused by 1/2 rod at given speed, or $\frac{66}{2}$ = 2033 gallons. At the intersection of 2033 gallons and 130 ft. piston-speed find the nearest diameter given on the diagram, which is 20 inches. Then 20-inch diameter delivering 2033 gallons = 127 ft. piston-speed. On the line representing this speed find the intersection of the length of stroke desired, and read the number of revolutions from the proper scale.

In the above case, if the stroke is 30 inches, the number of revolutions will be 25.5.

COAL REQUIRED IN PUMPING ONE MILLION GALLONS OF WATER.

Diagram No. 30.

This diagram is designed to give the weight of coal consumed in pumping-plants of different duties to raise one million gallons of water to various heads from 0 to 250. The head in feet is given on the horizontal scale. The duty is given by oblique lines from the upper left-hand corner. The weight of coal in net tons of 2000 lbs. is given on the vertical scale on the left, and the same in gross tons of 2240 lbs. on the right.

The following examples will be sufficient explanation of the use of the diagram.

Example 1: What weight of coal will be consumed in a pumping-plant capable of a duty of 60 million ft.-lbs. per 100 lbs. of coal to raise one million gallons of water against 150 ft. head?

At the intersection of 60 millions duty and 150 ft. head find weight of coal = 1.04 net tons or 0.92 gross tons.

Example 2: What is the duty of a pumping-plant that raises one million gallons of water for 2.3 net tons of coal consumed against a total head of 200 ft.? At intersection of 2.3 tons and 200 ft. head find duty = 36 millions approximately.

If problems of higher duties than 100 million are to be solved, divide the duty by some number, find answer for the quotient, and divide weight of coal found by same number.

Example: How much coal is required to pump one million gallons against a head of 175 feet with machinery capable of a duty of 125 million? Divide 125 million by 5 = 25 million; this with head of 175 will require 2.60 gross tons; this divided by 5 = .52 tons, answer for 125 million duty.

Reverse the process to find duty when coal is given.

Table No. 1.

TABLE OF U. S. GALLONS PER MINUTE AND THEIR EQUIVALENTS.

Gallons per Minute.	Gallons per 24 Hours.	Cubic Feet per Second.	Gallons per Minute.	Gallons per 24 Hours.	Cubic Feet per Second.
1	1440	0.002	350	504000	0.780
10	14400	.022	360	518400	.802
20	28800	.044	370	532800	.825
30	43200	.067	380	547200	.847
40	57600	.089	390	561600	.869
50	72000	.121	400	576000	.892
60	86400	.134	410	590400	.914
70	100800	.156	420	604800	.936
80	115200	.178	430	619200	.958
90	129600	.200	440	633600	.981
100	144000	.223	450	648000	1.003
110	158400	.245	460	662400	1.025
120	172800	.268	470	676800	1.048
130	187200	.290	480	691200	1.069
140	201600	.312	490	705600	1.091
150	216000	.335	500	720000	1.112
160	230400	.357	510	734400	1.136
170	244800	.380	520	748800	1.159
180	259200	.401	530	763200	1.181
190	273600	.424	540	777600	1.202
200	288000	.446	550	792000	1.222
210	302400	.468	560	806400	1.248
220	316800	.490	570	820800	1.269
230	331200	.513	580	835200	1.291
240	345600	.535	590	849600	1.312
250	360000	.557	600	864000	1.337
260	374400	.579	610	878400	1.359
270	388800	.601	620	892800	1.381
280	403200	.624	630	907200	1.402
290	417600	.647	640	921600	1.426
300	432000	.669	650	936000	1.449
310	446400	.691	660	950400	1.470
320	460800	.713	670	964800	1.492
330	475200	.736	680	979200	1.515
340	489600	.758	690	993600	1.538

U. S. GALLONS AND THEIR EQUIVALENTS.—*Continued.*

Gallons per Minute.	Gallons per 24 Hours.	Cubic Feet per Second.	Gallons per Minute.	Gallons per 24 Hours.	Cubic Feet per Second.
700	1008000	1.559	1250	1800000	2.785
710	1022400	1.581	1300	1872000	2.893
720	1036800	1.602	1350	1944000	3.009
730	1051200	1.627	1400	2016000	3.119
740	1065600	1.649	1450	2088000	3.230
750	1080000	1.671	1500	2160000	3.341
760	1094400	1.692	1550	2232000	3.453
770	1108800	1.715	1600	2304000	3.562
780	1123200	1.738	1650	2376000	3.676
790	1137600	1.760	1700	2448000	3.785
800	1152000	1.782	1750	2520000	3.899
810	1166400	1.802	1800	2592000	4.010
820	1180800	1.827	1850	2664000	4.121
830	1195200	1.849	1900	2736000	4.233
840	1209600	1.871	1950	2808000	4.344
850	1224000	1.892	2000	2880000	4.456
860	1238400	1.918	2050	2952000	4.567
870	1252800	1.936	2100	3024000	4.683
880	1267200	1.960	2150	3096000	4.790
890	1281600	1.982	2200	3168000	4.901
900	1296000	2.005	2250	3240000	5.013
910	1310400	2.027	2300	3312000	5.125
920	1324800	2.048	2350	3384000	5.235
930	1339200	2.073	2400	3456000	5.347
940	1353600	2.093	2450	3528000	5.458
950	1368000	2.114	2500	3600000	5.570
960	1382400	2.138	2550	3672000	5.681
970	1396800	2.161	2600	3744000	5.792
980	1411200	2.181	2650	3816000	5.904
990	1425600	2.202	2700	3888000	6.015
1000	1440000	2.228	2750	3960000	6.127
1050	1512000	2.339	2800	4032000	6.245
1100	1584000	2.450	2850	4104000	6.349
1150	1656000	2.562	2900	4176000	6.464
1200	1728000	2.672	2950	4248000	6.573
			3000	4320000	6.684

TABLE No. 2.

GIVING DISCHARGE OF CIRCULAR PIPES RUNNING FULL IN GALLONS PER MINUTE FOR DIFFERENT VELOCITIES.

Velocity in Feet per Second.	Diameter of Pipe in Inches.					
	4″	6″	8″	10″	12″	14″
.2	8	18	31	49	70	96
.4	16	35	63	98	141	192
.6	24	53	94	147	211	288
.8	31	70	125	196	282	384
1.0	39	88	157	245	352	480
.2	47	106	188	294	422	576
.4	55	123	219	342	493	672
.6	63	141	250	391	563	768
.8	70	159	282	441	634	864
2.0	78	176	313	489	704	960
.2	86	194	345	538	774	1056
.4	94	211	376	588	845	1152
.6	102	229	407	637	915	1248
.8	109	247	438	686	986	1344
3.0	117	264	470	734	1056	1440
.2	125	282	502	783	1126	1536
.4	133	300	532	832	1197	1632
.6	141	317	564	882	1267	1728
.8	149	335	596	930	1338	1824
4.0	156	352	627	979	1408	1920
.2	164	370	658	1027	1478	2016
.4	172	388	689	1076	1549	2112
.6	180	405	721	1125	1619	2208
.8	188	423	752	1174	1689	2304
5.0	196	441	783	1223	1760	2400
.5	215	485	861	1345	1936	2640
6.0	235	529	940	1468	2112	2880
.5	254	573	1017	1590	2288	3120
7.0	274	617	1096	1712	2464	3360
.5	293	661	1173	1835	2640	3600
8.0	313	705	1252	1957	2816	3840
.5	332	749	1330	2079	2992	4080
9.0	352	793	1410	2201	3168	4320
.5	372	837	1489	2324	3344	4560
10.0	391	881	1567	2446	3520	4800

DISCHARGE OF CIRCULAR PIPES FOR DIFFERENT VELOCITIES.—*Continued*.

Velocity in Feet per Second.	Diameter of Pipe in Inches.					
	15″	16″	18″	20″	24″	30″
.2	110	125	159	196	282	440
.4	220	251	318	391	564	880
.6	330	376	476	587	846	1320
.8	441	501	635	782	1128	1760
1.0	551	627	794	978	1410	2200
.2	661	752	953	1174	1692	2640
.4	771	877	1112	1369	1974	3080
.6	882	1003	1270	1565	2256	3520
.8	993	1128	1429	1760	2538	3960
2.0	1103	1253	1588	1956	2820	4400
.2	1213	1379	1747	2152	3102	4840
.4	1323	1504	1906	2347	3384	5280
.6	1434	1629	2064	2543	3666	5720
.8	1544	1755	2223	2738	3948	6160
3.0	1653	1880	2382	2934	4230	6600
.2	1763	2005	2541	3130	4512	7040
.4	1873	2131	2700	3325	4794	7480
.6	1984	2256	2858	3521	5076	7920
.8	2094	2381	3017	3716	5358	8360
4.0	2204	2507	3176	3912	5640	8800
.2	2314	2632	3335	4108	5922	9240
.4	2424	2757	3494	4302	6204	9680
.6	2535	2883	3652	4499	6486	10120
.8	2645	3008	3811	4694	6768	10560
5.0	2755	3133	3970	4890	7050	11000
.5	3031	3447	4367	5379	7755	12100
6.0	3306	3760	4764	5868	8460	13200
.5	3582	4073	5161	6357	9165	14300
7.0	3857	4387	5558	6846	9870	15400
.5	4133	4700	5955	7335	10575	16500
8.0	4408	5013	6352	7824	11280	17600
.5	4684	5326	6749	8313	11985	18700
9.0	4959	5641	7146	8802	12690	19800
.5	5235	5954	7543	9291	13395	20900
10.0	5510	6267	7940	9780	14100	22000

DISCHARGE OF CIRCULAR PIPES FOR DIFFERENT VELOCITIES.—*Continued*

Velocity in Feet per Second.	Diameter of Pipe in Inches.				
	36″	42″	48″	54″	60″
.2	631	863	1128	1427	1756
.4	1262	1727	2256	2854	3512
.6	1893	2590	3384	4281	5269
.8	2524	3454	4512	5708	7025
1.0	3155	4317	5639	7136	8781
.2	3786	5180	6767	8564	10537
.4	4417	6044	7895	9990	12293
.6	5048	6907	9023	11417	14049
.8	5679	7771	10151	12844	15807
2.0	6310	8634	11279	14272	17562
.2	6941	9497	12407	15699	19319
.4	7572	10361	13535	17126	21075
.6	8203	11224	14663	18553	22831
.8	8834	12088	15791	19981	24588
3.0	9465	12951	16919	21408	26344
.2	10096	13814	18046	22835	28100
.4	10727	14678	19175	24262	29856
.6	11358	15541	20303	25689	31613
.8	11989	16405	21431	27116	33369
4.0	12620	17268	22558	28544	35125
.2	13251	18131	23686	29971	36881
.4	13882	18995	24814	31398	38637
.6	14513	19858	25942	32825	40394
.8	15144	20722	27070	34252	42150
5.0	15775	21585	28198	35679	43906
.5	17353	23744	31018	39247	48297
6.0	18930	25902	33838	42815	52687
.5	20507	28060	36657	46383	57078
7.0	22085	30219	39474	49951	61468
.5	23662	32378	42297	53519	65858
8.0	25240	34536	45117	57087	70249
.5	26817	36694	47937	60655	74639
9.0	28395	38853	50756	64223	79030
.5	29973	41012	53576	67791	83420
10.0	31550	43170	56396	71359	87810

Table No. 3.
TABLE OF 11/6 AND 6/11 POWERS.

Number.	11/6 Power.	6/11 Power.	Number.	11/6 Power.	6/11 Power.	Number.	11/6 Power.	6/11 Power.
0.50	0.28	0.685	1.12	1.23	1.06	1.74	2.77	1.35
.52	.30	.70	1.14	1.27	1.07	1.76	2.83	1.36
.54	.32	.715	1.16	1.31	1.08	1.78	2.89	1.37
.56	.35	.73	1.18	1.36	1.09	1.80	2.95	1.38
.58	.37	.744	1.20	1.40	1.10	1.82	3.01	1.38
.60	.39	.758	1.22	1.44	1.11	1.84	3.07	1.39
.62	.42	.77	1.24	1.49	1.12	1.86	3.13	1.40
.64	.44	.785	1.26	1.53	1.13	1.88	3.19	1.41
.66	.47	.798	1.28	1.58	1.14	1.90	3.26	1.42
.68	.49	.81	1.30	1.62	1.15	1.92	3.32	1.43
.70	.52	.824	1.32	1.67	1.16	1.94	3.38	1.43
.72	.55	.835	1.34	1.71	1.17	1.96	3.45	1.44
.74	.58	.848	1.36	1.76	1.18	1.98	3.52	1.45
.76	.61	.86	1.38	1.81	1.19	2.	3.56	1.46
.78	.64	.873	1.40	1.86	1.20	2.50	5.40	1.65
.80	.67	.885	1.42	1.91	1.21	3.00	7.55	1.82
.82	.70	.899	1.44	1.96	1.22	3.50	10.	1.98
.84	.73	.91	1.46	2.01	1.23	4.00	12.75	2.13
.86	.76	.92	1.48	2.06	1.24	4.50	15.85	2.27
.88	.79	.932	1.50	2.11	1.25	5.00	19.20	2.40
.90	.82	.944	1.52	2.16	1.25	5.50	22.90	2.53
.92	.86	.955	1.54	2.21	1.26	6.00	26.90	2.65
.94	.89	.966	1.56	2.27	1.27	6.50	31.20	2.77
.96	.93	.977	1.58	2.32	1.28	7.00	35.70	2.88
.98	.97	.988	1.60	2.37	1.29	7.50	40.70	2.99
1.00	1.	1.	1.62	2.43	1.30	8.00	45.60	3.10
1.02	1.04	1.01	1.64	2.49	1.31	8.50	51.00	3.20
1.04	1.07	1.02	1.66	2.54	1.32	9.00	56.70	3.30
1.06	1.11	1.03	1.68	2.60	1.33	9.50	62.70	3.40
1.08	1.15	1.04	1.70	2.65	1.33	10.00	68.00	3.50
1.10	1.19	1.05	1.72	2.71	1.34			

Table No. 4.

COEFFICIENTS FOR LOSS OF HEAD BY FRICTION IN OLD PIPE-LINES.

Velocity, Feet per Second.	Age of Pipe in Years.							
	5	10	15	20	25	30	40	50
1	1.10	1.21	1.33	1.46	1.58	1.71	1.96	2.23
2	1.14	1.30	1.47	1.65	1.82	2.00	2.37	2.74
3	1.17	1.37	1.57	1.80	2.00	2.22	2.69	3.13
4	1.20	1.43	1.66	1.93	2.16	2.41	2.96	3.45
5	1.22	1.48	1.74	2.04	2.30	2.57	3.20	3.73
6	1.24	1.52	1.81	2.13	2.42	2.72	3.41	3.99
7	1.26	1.56	1.88	2.22	2.53	2.86	3.61	4.22
8	1.28	1.60	1.93	2.31	2.63	2.99	3.79	4.44
9	1.30	1.64	1.99	2.39	2.73	3.11	3.97	4.65
10	1.32	1.67	2.04	2.46	2.82	3.22	4.13	4.85

Table No. 5

COEFFICIENTS OF DISCHARGE IN OLD PIPE-LINES.

Velocity, Feet per Second.	Age of Pipe in Years.							
	5	10	15	20	25	30	40	50
1	0.95	0.91	0.86	0.82	0.79	0.77	0.72	0.68
2	.93	.87	.82	.78	.75	.72	.66	.62
3	.92	.85	.79	.75	.71	.68	.63	.59
4	.91	.84	.78	.73	.69	.66	.60	.56
5	.90	.83	.77	.71	.67	.64	.58	.54
6	.89	.82	.75	.69	.66	.62	.57	.53
7	.89	.80	.74	.68	.65	.61	.56	.52
8	.88	.79	.72	.67	.63	.60	.54	.51
9	.88	.79	.72	.66	.63	.59	.53	.50
10	.87	.78	.71	.65	.62	.58	.53	.49

TABLES. 67

Table No. 6.

GIVING CORRECTIONS IN FEET TO BE ADDED TO HEAD ON THIN-CREST WEIRS FOR VELOCITY OF APPROACH.

WITHOUT END CONTRACTIONS.

Depth of Bottom of Channel below Crest.

Velocity of Approach.	Depth below Crest = .50.				Depth below Crest = 1.			Depth below Crest = 2.50.			
	Head on Crest.				Head on Crest.			Head on Crest.			
	.20	.30 to .50	.40 to .60	.80 to 1.00				.50	1.00	1.50	2.00
.2	.001	.001	.001	.001				.001	.001	.001	.001
.3	.002	.002	.002	.002				.002	.002	.002	.002
.4	.004	.004	.004	.004				.004	.004	.003	.003
.5	.007	.006	.007	.006				.006	.006	.005	.005
.6	.01	.009	.01	.009				.008	.008	.008	.007
.7	....	.012	.013	.012				.011	.011	.011	.01
.8	....	.015	.017	.016				.015	.014	.014	.013
.9	....	....	.022	.021				.019	.018	.017	.017
1.0	....	....	.027	.026				.023	.022	.021	.021
1.1	....	....	.033	.031				.028	.027	.026	.025
1.2	....	....	.039	.037				.033	.032	.031	.030
1.3	....	....	....	.043				.039	.038	.036	.035
1.4	....	....	....	.051				.045	.044	.042	.041
1.5	....	....	....	.058				.052	.05	.048	.047
1.6	....	....	....	....				.059	.057	.055	.053
1.7	....	....	....	....				....	.064	.062	.06
1.8	....	....	....	....				....	.072	.069	.067
1.9	....	....	....	....				....	.08	.077	.075
2.	....	....	....	....				....	....	....	.083

WITH END CONTRACTIONS.

.2	.001	.001	.001	.001				.001	.001	.001	.001
.3	.003	.003	.003	.003				.003	.003	.003	.003
.4	.006	.005	.006	.005				.005	.005	.005	.005
.5	.009	.008	.009	.008				.008	.008	.008	.007
.6	.013	.012	.013	.012				.011	.011	.011	.011
.7	.017	.016	.017	.017				.015	.015	.015	.014
.8	....	.021	.023	.022				.02	.02	.019	.019
.9	....	....	.029	.027				.026	.025	.024	.024
1.0	....	....	.036	.034				.032	.031	.03	.029
1.1	....	....	.043	.041				.038	.037	.036	.036
1.2	....	....	.051	.048				.046	.044	.043	.042
1.3	....	....	....	.057				.053	.052	.051	.049
1.4	....	....	....	.066				.062	.06	.059	.058
1.5	....	....	....	.076				.071	.069	.068	.065
1.6	....	....	....	....				.081	.079	.077	.075
1.7	....	....	....	....				....	.089	.087	.085
1.8	....	....	....	....				....	.099	.097	.094
1.9	....	....	....	....				....	.111	.108	.105
2.	....	....	....	....				....	....	....	.117

TABLE No. 7.
GIVING AREA AND VALUE OF r OF CIRCULAR CHANNELS RUNNING FULL.

Diameter in Ft. and In.	Area in Sq. Ft.	Values of r	Diameter in Ft. and In.	Area in Sq. Ft.	Values of r	Diameter in Ft. and In.	Area in Sq. Ft.	Values of r
1"	0.005	0.02	3' 5"	9.17	.85	6' 9"	35.78	1.69
2	.022	.04	6	9.62	.88	10	36.67	1.71
3	.049	.06	7	10.08	.90	11	37.57	1.73
4	.087	.08	8	10.56	.92	7'	38.48	1.75
5	.136	.10	9	11.04	.94	1	39.41	1.77
6	.196	.13	10	11.54	.96	2	40.34	1.79
7	.267	.15	11	12.05	.98	3	41.28	1.81
8	.349	.17	4'	12.57	1.00	4	42.24	1.83
9	.442	.19	1	13.10	1.02	5	43.20	1.85
10	.545	.21	2	13.64	1.04	6	44.18	1.88
11	.660	.23	3	14.19	1.06	7	45.17	1.90
1'	.785	.25	4	14.75	1.08	8	46.16	1.92
1	.921	.27	5	15.32	1.10	9	47.17	1.94
2	1.07	.30	6	15.90	1.13	10	48.19	1.96
3	1.23	.31	7	16.50	1.15	11	49.22	1.97
4	1.40	.33	8	17.10	1.17	8'	50.27	2.00
5	1.58	.35	9	17.72	1.19	1	51.32	2.02
6	1.77	.38	10	18.35	1.21	2	52.38	2.04
7	1.97	.40	11	18.99	1.23	3	53.46	2.06
8	2.18	.42	5'	19.63	1.25	4	54.54	2.08
9	2.41	.44	1	20.29	1.27	5	55.64	2.10
10	2.64	.46	2	20.97	1.29	6	56.75	2.13
11	2.89	.48	3	21.65	1.31	7	57.86	2.15
2'	3.14	.50	4	22.34	1.33	8	58.99	2.17
1	3.41	.52	5	23.04	1.35	9	60.13	2.19
2	3.69	.54	6	23.76	1.38	10	61.28	2.21
3	3.98	.56	7	24.48	1.40	11	62.44	2.23
4	4.28	.58	8	25.22	1.42	9'	63.62	2.25
5	4.59	.61	9	25.97	1.44	1	64.80	2.27
6	4.91	.63	10	26.73	1.46	2	66.	2.29
7	5.24	.65	11	27.49	1.48	3	67.20	2.31
8	5.59	.67	6'	28.27	1.50	4	68.42	2.33
9	5.94	.69	1	29.07	1.52	5	69.64	2.35
10	6.31	.71	2	29.87	1.54	6	70.88	2.38
11	6.68	.73	3	30.68	1.56	7	72.13	2.40
3'	7.07	.75	4	31.50	1.58	8	73.39	2.42
1	7.47	.77	5	32.34	1.60	9	74.66	2.44
2	7.88	.79	6	33.18	1.63	10	75.94	2.46
3	8.30	.81	7	34.04	1.65	11	77.24	2.48
4	8.73	.83	8	34.91	1.67	10'	78.54	2.50

TABLE No. 8.
GIVING AREA AND r IN RECTANGULAR CHANNELS.

Depth.	$W = 2$ Ft. Area.	r	$W = 4$ Ft. Area.	r	$W = 6$ Ft. Area.	r	$W = 8$ Ft. Area.	r	$W = 10$ Ft. Area.	r	$W = 15$ Ft. Area.	r	$W = 20$ Ft. Area.	r	$W = 25$ Ft. Area.	r
0.2	0.4	0.17	0.8	0.18	1.2	0.19	1.6	0.19	2	0.19	3	0.19	4	0.19	5	0.2
.4	.8	.28	1.6	.33	2.4	.35	3.2	.37	4	.37	6	.38	8	.38	10	.39
.6	1.2	.37	2.4	.46	3.6	.50	4.8	.53	6	.54	9	.56	12	.57	15	.57
.8	1.6	.45	3.2	.57	4.8	.63	6.4	.67	8	.69	12	.72	16	.74	20	.76
1	2	.50	4	.67	6	.75	8	.80	10	.83	15	.88	20	.91	25	.93
1.25	2.5	.55	5	.77	7.5	.88	10	.96	12.5	1.01	18.75	1.07	25	1.12	31.25	1.14
1.50	3	.60	6	.86	9	1	12	1.09	15	1.16	22.50	1.25	30	1.31	37.50	1.34
1.75	3.5	.63	7	.94	10.5	1.10	14	1.22	17.5	1.30	26.25	1.42	35	1.49	43.75	1.54
2	4	.67	8	1	12	1.20	16	1.33	20	1.43	30	1.58	40	1.67	50	1.72
2.25	4.5	.69	9	1.06	13.5	1.28	18	1.43	22.5	1.55	33.75	1.73	45	1.83	56.25	1.91
2.50	5	.71	10	1.11	15	1.36	20	1.53	25	1.67	37.50	1.87	50	2	62.50	2.08
2.75	5.5	.73	11	1.16	16.5	1.44	22	1.63	27.5	1.77	41.25	2.02	55	2.15	68.75	2.26
3	6	.75	12	1.20	18	1.50	24	1.71	30	1.88	45	2.14	60	2.30	75	2.42
3.50	7	.77	14	1.27	21	1.62	28	1.86	35	2.05	52.50	2.39	70	2.51	87.50	2.74
4	8	.80	16	1.33	24	1.72	32	2	40	2.22	60	2.61	80	2.87	100	3.03
4.50	9	.82	18	1.39	27	1.80	36	2.12	45	2.37	67.50	2.82	90	3.10	112.50	3.33
5	10	.83	20	1.43	30	1.88	40	2.23	50	2.54	75	3	100	3.33	125	3.59
6	12	.86	24	1.50	36	2	48	2.40	60	2.74	90	3.33	120	3.75	150	4.08
7	14	.88	28	1.56	42	2.11	56	2.56	70	2.93	105	3.63	140	4.13	175	4.52
8	16	.89	32	1.61	48	2.19	64	2.68	80	3.09	120	3.87	160	4.45	200	4.90
9	18	.90	36	1.64	54	2.26	72	2.77	90	3.22	135	4.10	180	4.73	225	5.26
10	20	.91	40	1.67	60	2.31	80	2.86	100	3.33	150	4.29	200	5	250	5.56

TABLE No. 9.

CHANNELS OF TRAPEZOIDAL SECTION. SIDE SLOPES 1 TO 1.

GIVING AREA AND VALUES OF r.

Depth.	$W = 1$ Ft.		$W = 2$ Ft.		$W = 3$ Ft.		$W = 4$ Ft.	
	Area.	r	Area.	r	Area.	r	Area.	r
0.5	0.75	0.31	1.25	0.37	1.75	0.40	2.25	0.42
.75	1.32	.42	2.07	.50	2.82	.55	3.57	.58
1	2	.52	3	.62	4	.69	5	.73
1.25	2.82	.62	4.07	.74	5.32	.81	6.57	.87
1.50	3.75	.72	5.25	.84	6.75	.93	8.25	1
1.75	4.81	.81	6.56	.94	8.31	1.05	10.06	1.12
2	6	.90	8	1.04	10	1.15	12	1.24
2.25	7.35	.99	9.60	1.14	11.85	1.26	14.10	1.36
2.50	8.75	1.08	11.25	1.24	13.75	1.37	16.25	1.47
2.75	10.30	1.17	13.05	1.33	15.80	1.47	18.55	1.57
3	12	1.27	15	1.43	18	1.57	21	1.68
3.50	15.75	1.45	19.25	1.62	22.75	1.77	26.25	1.89
4	20	1.63	24	1.80	28	1.96	32	2.09
4.50	24.75	1.80	29.25	1.99	33.75	2.14	38.25	2.29
5	30	1.98	35	2.17	40	2.33	45	2.48

Depth.	$W = 5$ Ft.		$W = 6$ Ft.		$W = 8$ Ft.		$W = 10$ Ft.	
	Area.	r	Area.	r	Area.	r	Area.	r
0.5	2.75	0.43	3.25	0.44	4.25	0.45	5.25	0.46
.75	4.32	.61	5.07	.63	6.57	.65	8.07	.67
1	6	.77	7	.79	9	.83	11	.86
1.25	7.82	.92	9.07	.95	11.57	1	14.07	1.04
1.50	9.75	1.05	11.25	1.11	14.25	1.16	17.25	1.21
1.75	11.81	1.18	13.56	1.24	17.06	1.32	20.56	1.38
2	14	1.31	16	1.37	20	1.47	24	1.53
2.25	16.35	1.43	18.60	1.50	23.10	1.61	27.60	1.68
2.50	18.75	1.55	21.25	1.63	26.25	1.74	31.25	1.83
2.75	21.30	1.67	24.05	1.75	29.55	1.87	35.05	1.97
3	24	1.78	27	1.88	33	2	39	2.11
3.50	29.75	2	33.25	2.09	40.25	2.25	47.25	2.38
4	36	2.21	40	2.31	48	2.48	56	2.63
4.50	42.75	2.41	47.25	2.52	56.25	2.71	65.25	2.87
5	50	2.61	55	2.73	65	2.94	75	3.10

TABLES. 71

TABLE No. 10.
CHANNELS OF TRAPEZOIDAL SECTION. SIDE SLOPES 2 TO 1.
GIVING AREA AND VALUES OF r.

Depth.	$W = 1$ Ft.		$W = 2$ Ft.		$W = 3$ Ft.		$W = 4$ Ft.	
	Area.	r	Area.	r	Area.	r	Area.	r
0.50	1	0.31	1.50	0.35	2	0.38	2.50	0.40
.75	1.88	.43	2.63	.49	3.38	.52	4.13	.56
1	3	.55	4	.62	5	.67	6	.71
1.25	4.37	.66	5.62	.74	6.87	.80	8.12	.84
1.50	6	.78	7.50	.86	9	.93	10.50	.98
1.75	7.87	.90	9.62	.98	11.37	1.05	13.12	1.11
2	10	1.01	12	1.10	14	1.17	16	1.24
2.25	12.40	1.12	14.65	1.21	16.90	1.29	19.15	1.36
2.50	15	1.23	17.50	1.33	20	1.41	22.50	1.48
2.75	17.85	1.34	20.60	1.44	23.35	1.53	26.10	1.60
3	21	1.45	24	1.56	27	1.64	30	1.72
3.50	28	1.68	31.50	1.79	35	1.87	38.50	1.96
4	36	1.90	40	2.01	44	2.11	48	2.19
4.50	45	2.13	49.50	2.24	54	2.34	58.50	2.42
5	55	2.36	60	2.46	65	2.56	70	2.66

Depth.	$W = 5$ Ft.		$W = 6$ Ft.		$W = 8$ Ft.		$W = 10$ Ft.	
	Area.	r	Area.	r	Area.	r	Area.	r
0.50	3	0.41	3.50	0.42	4.50	0.44	5.50	0.45
.75	4.88	.58	5.63	.60	7.13	.63	8.63	.65
1	7	.74	8	.76	10	.80	12	.83
1.25	9.37	.88	10.62	.92	13.12	.96	15.62	1
1.50	12	1.02	13.50	1.06	16.50	1.12	19.50	1.17
1.75	14.87	1.15	16.62	1.20	20.12	1.27	23.62	1.33
2	18	1.29	20	1.34	24	1.42	28	1.48
2.25	21.40	1.42	23.65	1.47	28.15	1.56	32.65	1.63
2.50	25	1.55	27.50	1.60	32.50	1.69	37.50	1.77
2.75	28.85	1.67	31.60	1.73	37.10	1.83	42.60	1.91
3	33	1.79	36	1.85	42	1.96	48	2.05
3.50	42	2.03	45.50	2.10	52.50	2.22	59.50	2.32
4	52	2.27	56	2.34	64	2.47	72	2.58
4.50	63	2.51	67.50	2.59	76.50	2.72	85.50	2.84
5	75	2.74	80	2.82	90	2.97	100	3.09

TABLE No. 11.

CHANNELS OF TRAPEZOIDAL SECTION. SIDE SLOPES 3 TO 1.
GIVING AREA AND VALUES OF r.

Depth.	$W = 1$ Ft.		$W = 2$ Ft.		$W = 3$ Ft.		$W = 4$ Ft.	
	Area.	r	Area.	r	Area.	r	Area.	r
0.50	1.25	0.30	1.75	0.34	2.25	0.37	2.75	0.39
.75	2.44	.43	3.19	.47	3.94	.57	4.69	.54
1	4	.55	5	.60	6	.65	7	.68
1.25	5.95	.67	7.20	.73	8.45	.78	9.70	.82
1.50	8.25	.79	9.75	.85	11.25	.90	12.75	.95
1.75	10.92	.91	12.67	.97	14.42	1.02	16.17	1.07
2	14	1.03	16	1.10	18	1.15	20	1.20
2.25	17.50	1.15	19.75	1.21	22	1.27	24.25	1.33
2.50	21.25	1.26	23.75	1.33	26.25	1.39	28.75	1.45
2.75	25.50	1.38	28.25	1.45	31	1.52	33.75	1.57
3	30	1.50	33	1.57	36	1.64	39	1.70
3.50	40.25	1.74	43.75	1.81	47.25	1.88	50.75	1.94
4	52	1.97	56	2.05	60	2.12	64	2.18
4.50	65.30	2.20	69.80	2.28	74.30	2.35	78.80	2.42
5	30	2.45	85	2.53	90	2.60	95	2.67

Depth.	$W = 5$ Ft.		$W = 6$ Ft.		$W = 8$ Ft.		$W = 10$ Ft.	
	Area.	r	Area.	r	Area.	r	Area.	r
0.50	3.25	0.40	3.75	0.41	4.75	0.43	5.75	0.44
.75	5.44	.55	6.19	.58	7.69	.60	9.19	.62
1	8	.70	9	.73	11	.77	13	.80
1.25	10.95	.85	12.20	.88	14.70	.93	17.20	.96
1.50	14.25	.99	15.75	1.02	18.75	1.07	21.75	1.12
1.75	17.92	1.12	19.67	1.15	23.17	1.21	26.67	1.27
2	22	1.25	24	1.29	28	1.36	32	1.42
2.25	26.50	1.38	28.75	1.42	33.25	1.49	37.75	1.56
2.50	31.25	1.50	33.75	1.55	38.75	1.63	43.75	1.70
2.75	36.50	1.63	39.25	1.68	44.75	1.76	50.25	1.83
3	42	1.75	45	1.80	51	1.89	57	1.97
3.50	54.25	2	57.75	2.05	64.75	2.15	71.75	2.24
4	68	2.24	72	2.30	80	2.40	88	2.49
4.50	83.30	2.48	87.80	2.59	96.80	2.64	105.80	2.74
5	100	2.73	105	2.80	115	2.91	125	3.01

Table No. 12.

CONVERSION TABLE OF PRESSURE IN POUNDS AND HEAD IN FEET.

Pressure in Pounds.	Head in Feet.	Pressure in Pounds.	Head in Feet.	Pressure in Pounds.	Head in Feet.	Pressure in Pounds.	Head in Feet.	Pressure in Pounds.	Head in Feet.
1	2.3	41	94.6	66	152.3	91	210	116	267.7
5	11.5	42	96.9	67	154.6	92	212.3	117	270
10	23.1	43	99.2	68	156.9	93	214.6	118	272.3
15	34.6	44	101.5	69	159.2	94	216.9	119	274.6
20	46.2	45	103.9	70	161.6	95	219.2	120	276.9
21	48.5	46	106.2	71	163.9	96	221.5	121	279.2
22	50.8	47	108.5	72	166.2	97	223.9	122	281.6
23	53.1	48	110.8	73	168.5	98	226.2	123	283.9
24	55.4	49	113.1	74	170.8	99	228.5	124	286.2
25	57.7	50	115.4	75	173.1	100	230.8	125	288.5
26	60	51	117.7	76	175.4	101	233.1	130	300
27	62.3	52	120	77	177.7	102	235.4	135	311.6
28	64.6	53	122.3	78	180	103	237.7	140	323.1
29	66.9	54	124.6	79	182.3	104	240	145	334.6
30	69.2	55	126.9	80	184.6	105	242.3	150	346.2
31	71.5	56	129.2	81	186.9	106	244.6	155	357.7
32	73.9	57	131.5	82	189.2	107	246.9	160	369.3
33	76.2	58	133.9	83	191.6	108	249.2	165	380.8
34	78.5	59	136.2	84	193.9	109	251.6	170	392.3
35	80.8	60	138.5	85	196.2	110	253.9	175	403.9
36	83.1	61	140.8	86	198.5	111	256.2	180	415.4
37	85.4	62	143.1	87	200.8	112	258.5	185	427
38	87.7	63	145.4	88	203.1	113	260.8	190	438.5
39	90	64	147.7	89	205.4	114	263.1	195	450.1
40	92.3	65	150	90	207.7	115	265.4	200	461.6

74 GRAPHICAL SOLUTION OF HYDRAULIC PROBLEMS.

TABLE No. 13.

GIVING HEAD IN FEET REQUIRED AT HYDRANT FOR FIRE-STREAMS.

(Compiled from the tables of JOHN R. FREEMAN, Hydraulic Engineer.)

Head Indicated by Gauge at the End of Play-pipe.	Height of Effective Fire-stream in Moderate Wind in Feet.	Gallons per Minute.	Head Required at Hydrant to Maintain Head at Base of Play-pipe as in First Column while Stream is Running through Ordinary Best Quality Rubber-lined Hose 2½″ Diameter.					
			Length of Hose.					
			50 Ft.	100 Ft.	200 Ft.	300 Ft.	400 Ft.	500 Ft.
1-INCH SMOOTH NOZZLE.								
Feeble streams { 35	26	114	39	44	51	58	65	69
46	35	132	55	58	67	76	85	95
Fair .... { 58	43	147	67	72	83	95	106	117
69	51	161	78	85	99	113	127	140
Good .... { 81	58	174	92	101	117	131	147	164
92	64	186	106	115	134	152	168	187
104	69	198	120	129	150	170	191	210
Excellent .... { 115	73	208	131	143	166	189	212	235
127	76	218	145	159	182	207	233	258
138	79	228	159	173	200	226	254	281
Maximum Limit as a Fair Fire-stream to Per Cent Higher.								
Unusually strong { 150	82	237	173	187	216	247	275	305
161	85	246	184	201	233	265	296	325
173	87	255	198	217	249	284	321	348
184	89	263	212	231	265	302	343	372

1¼-INCH SMOOTH NOZZLE.

	Maximum Limit as a Fair Fire-stream to Per Cent Higher.								
Feeble streams	35 / 46	27 / 36	146 / 168	44 / 58	48 / 65	60 / 78	71 / 95	81 / 108	92 / 124
Fair	58 / 69	44 / 52	188 / 206	71 / 85	81 / 97	99 / 120	118 / 141	136 / 164	154 / 184
Good	81 / 92 / 104	59 / 65 / 70	222 / 238 / 252	90 / 115 / 124	113 / 129 / 145	138 / 159 / 178	164 / 187 / 212	189 / 217 / 244	217 / 247 / 277
Excellent	115 / 127 / 138	75 / 80 / 83	266 / 279 / 291	143 / 157 / 170	161 / 178 / 194	198 / 219 / 237	235 / 258 / 281	272 / 300 / 325	309 / 339 / 369
Unusually strong	150 / 161 / 173 / 184	86 / 88 / 90 / 92	303 / 314 / 325 / 336	186 / 201 / 215 / 228	210 / 226 / 242 / 258	258 / 276 / 298 / 318	304 / 329 / 353 / 376	352 / 380 / 408 / 433	401 / 431 / 463 / 493

1⅜-INCH SMOOTH NOZZLE.

	Maximum Limit as a Fair Fire-stream to Per Cent Higher.								
Feeble streams	35 / 46	28 / 37	181 / 209	48 / 62	55 / 74	74 / 97	90 / 120	108 / 143	125 / 166
Fair	58 / 69	46 / 53	234 / 256	78 / 95	92 / 113	122 / 145	150 / 180	178 / 215	208 / 249
Good	81 / 92 / 104	60 / 67 / 72	277 / 296 / 314	111 / 127 / 143	131 / 150 / 168	171 / 194 / 219	210 / 240 / 270	251 / 286 / 323	290 / 332 / 373
Excellent	115 / 127 / 138	77 / 81 / 85	331 / 347 / 363	157 / 173 / 189	187 / 205 / 224	244 / 267 / 293	300 / 330 / 360	357 / 392 / 428	415 / 456 / 493
Unusually strong	150 / 161 / 173 / 184	88 / 91 / 93 / 95	377 / 392 / 405 / 419	205 / 221 / 237 / 253	242 / 261 / 279 / 297	316 / 341 / 364 / 389	390 / 420 / 450 / 480	461 / 500 / 535 / 571	540 / 580

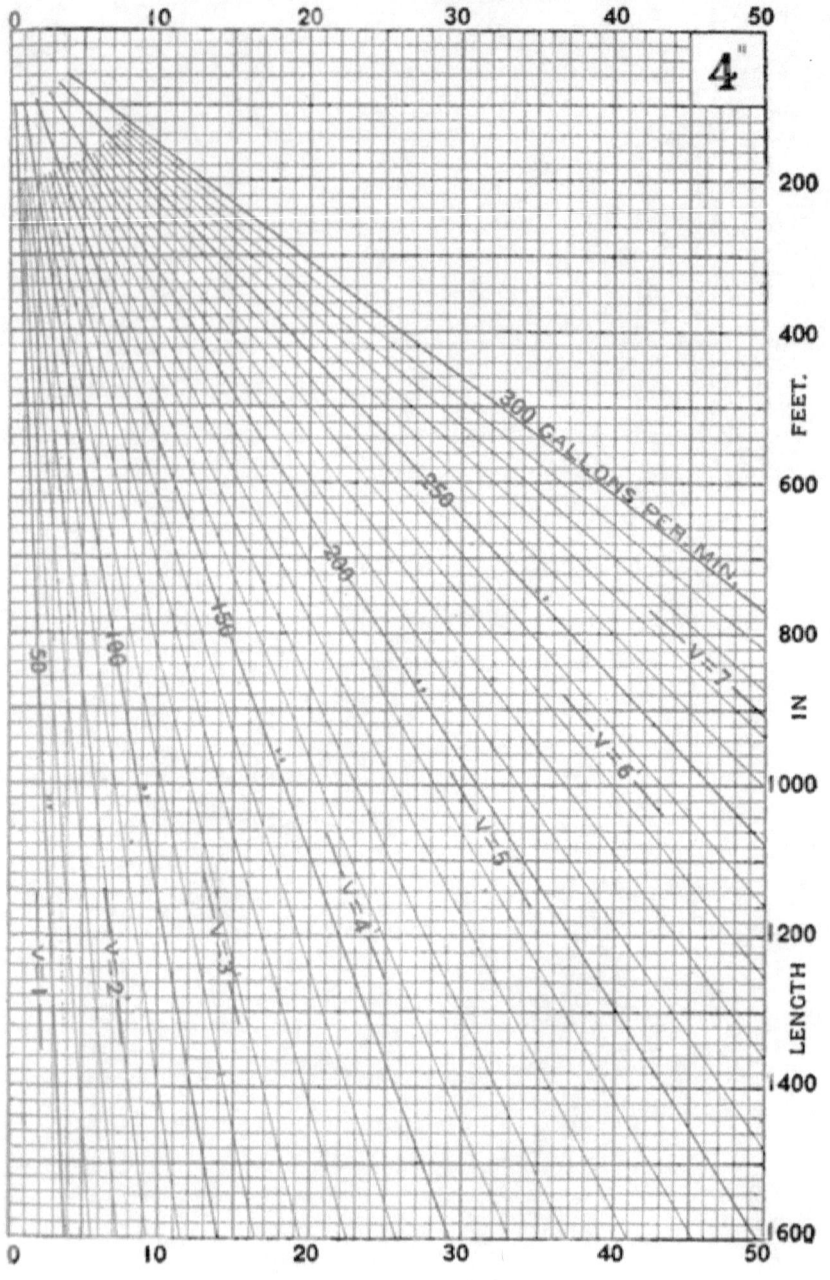

FLOW OF WATER IN LONG PIPES.

DIAGRAM No. 2

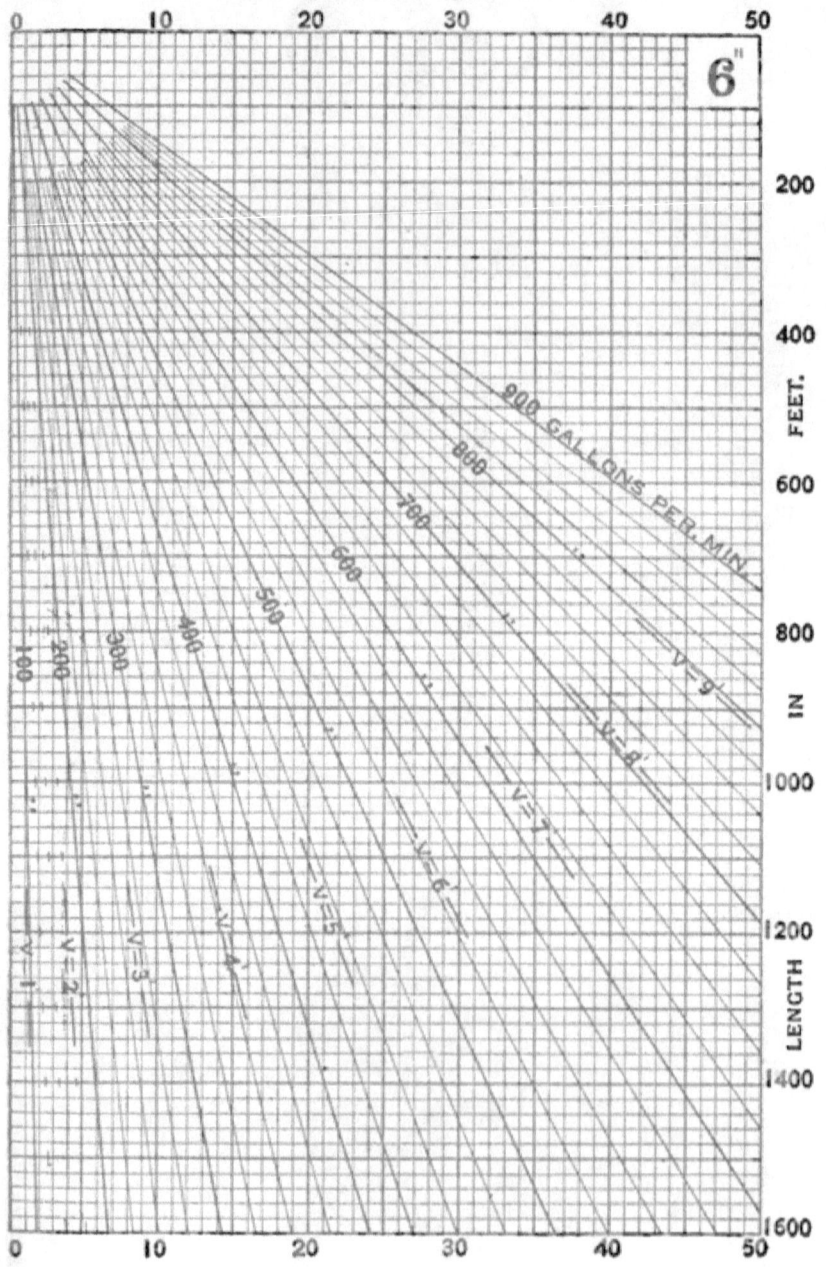

FLOW OF WATER IN LONG PIPES.

DIAGRAM No. 3

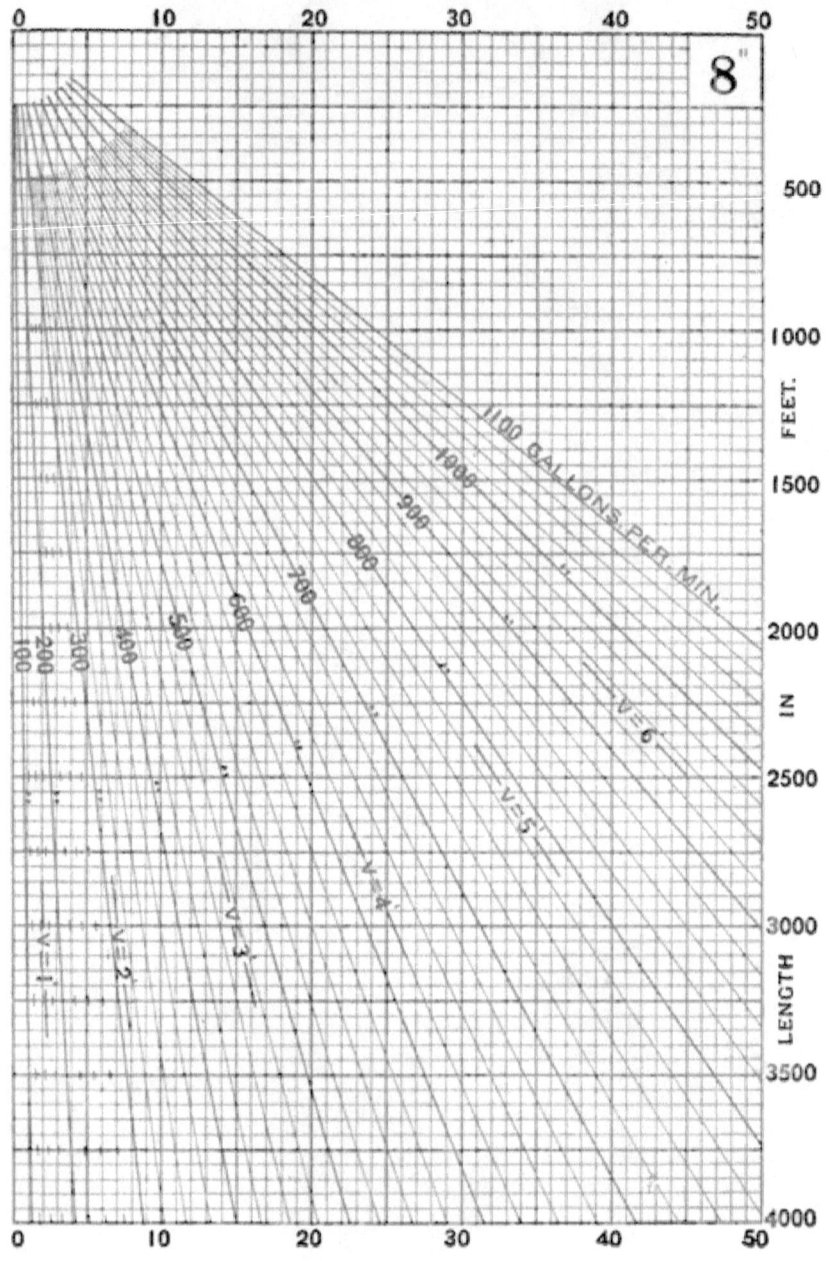

FLOW OF WATER IN LONG PIPES.

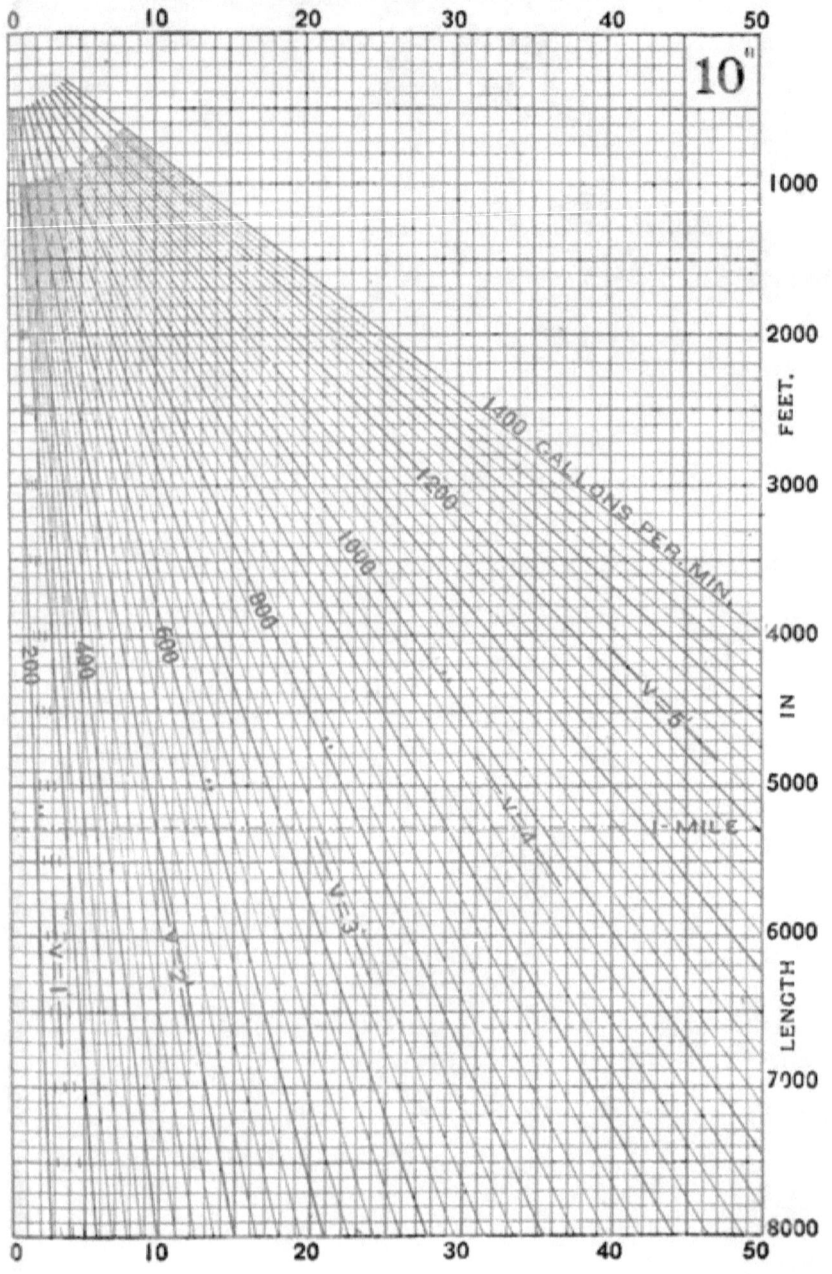

FLOW OF WATER IN LONG PIPES.
DIAGRAM No. 5

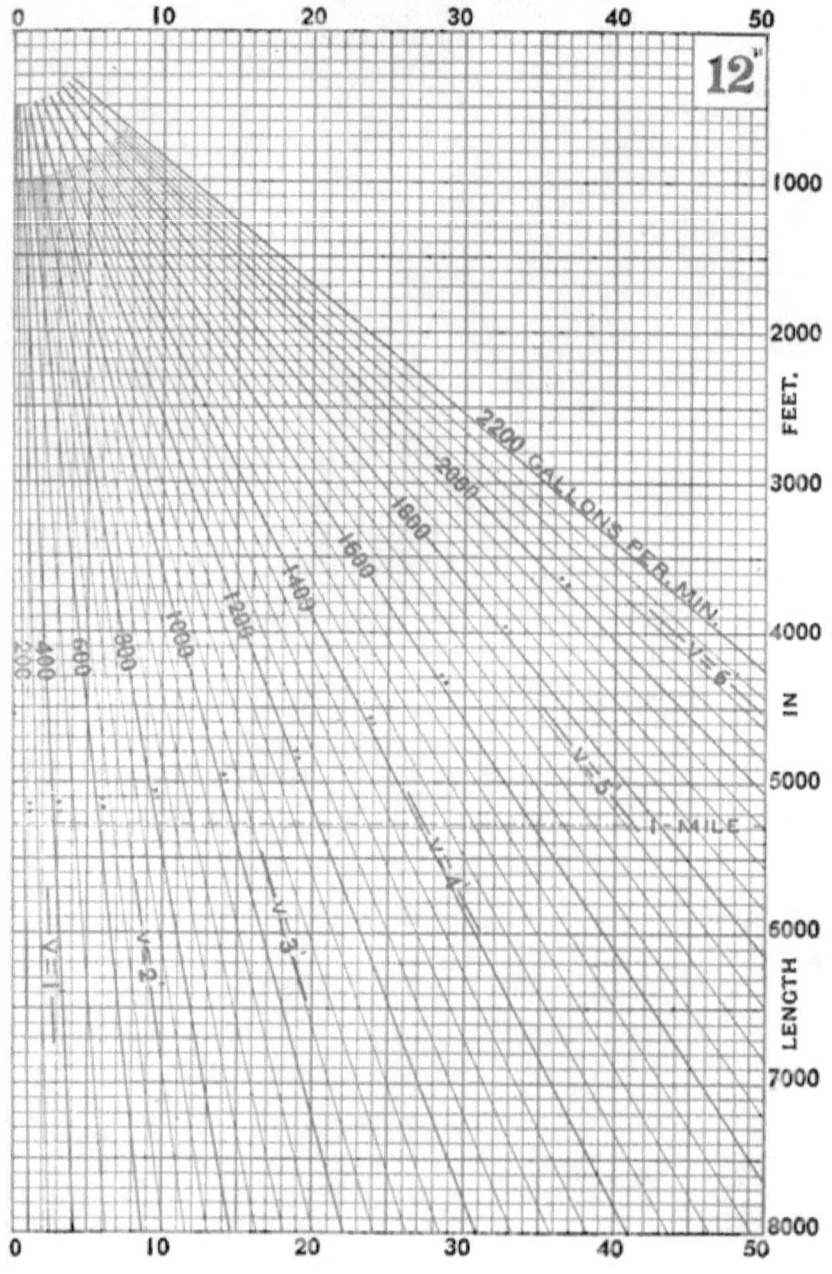

FLOW OF WATER IN LONG PIPES.
DIAGRAM No. 6

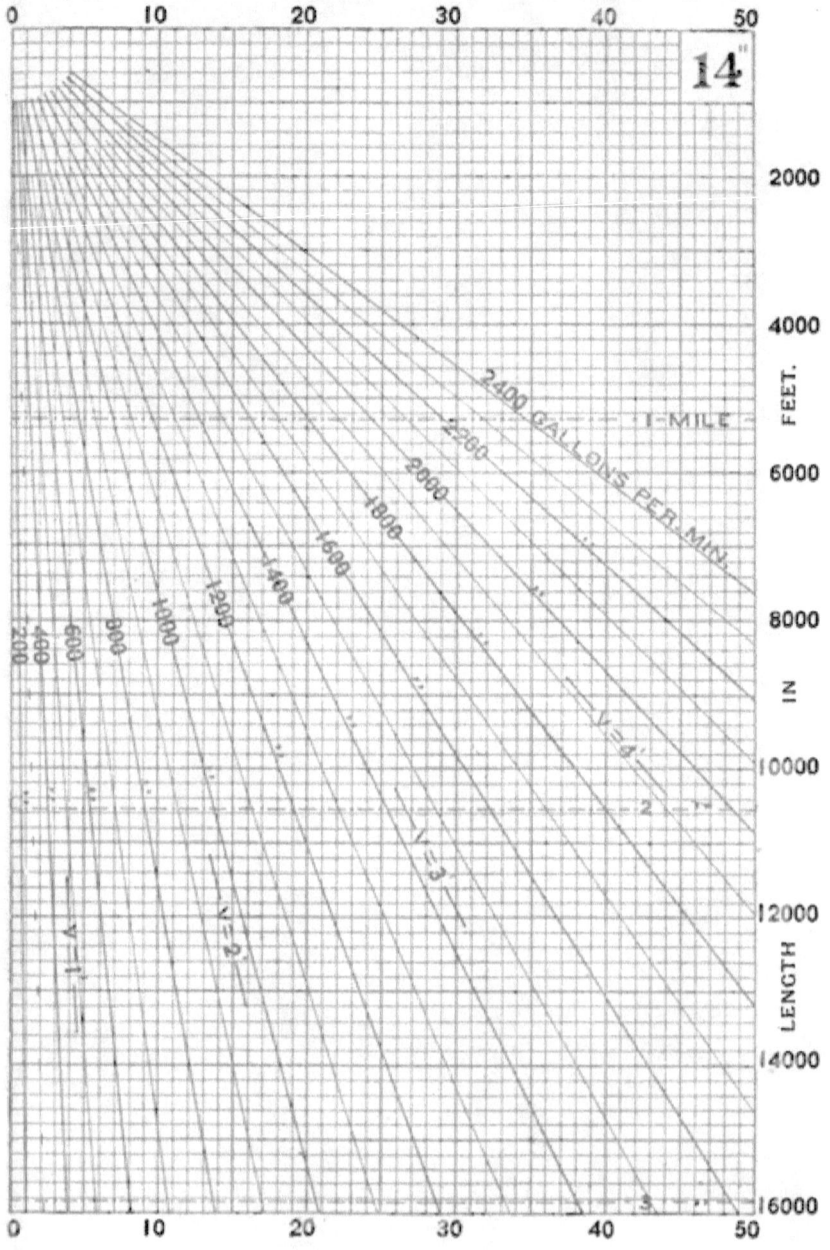

FLOW OF WATER IN LONG PIPES.

DIAGRAM No. 7

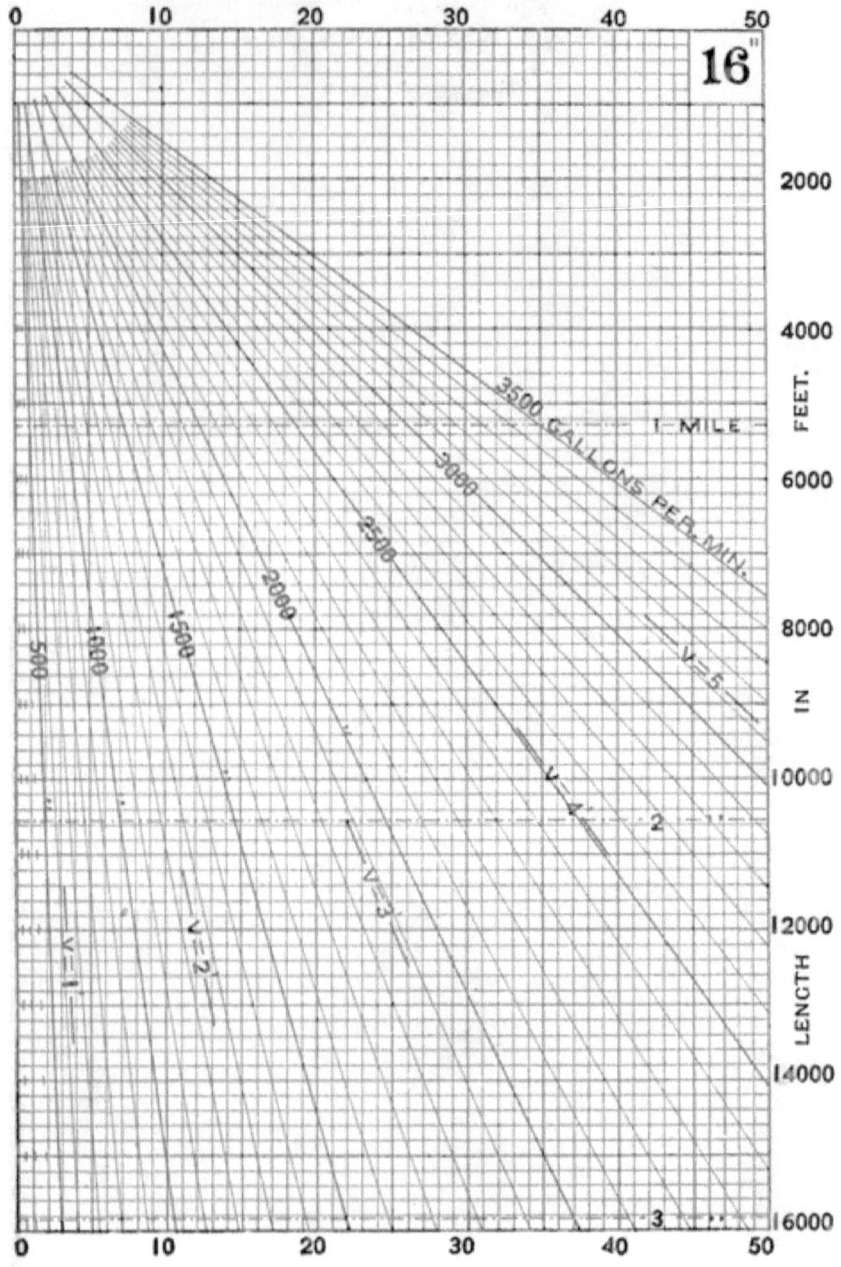

FLOW OF WATER IN LONG PIPES.

DIAGRAM No. 8

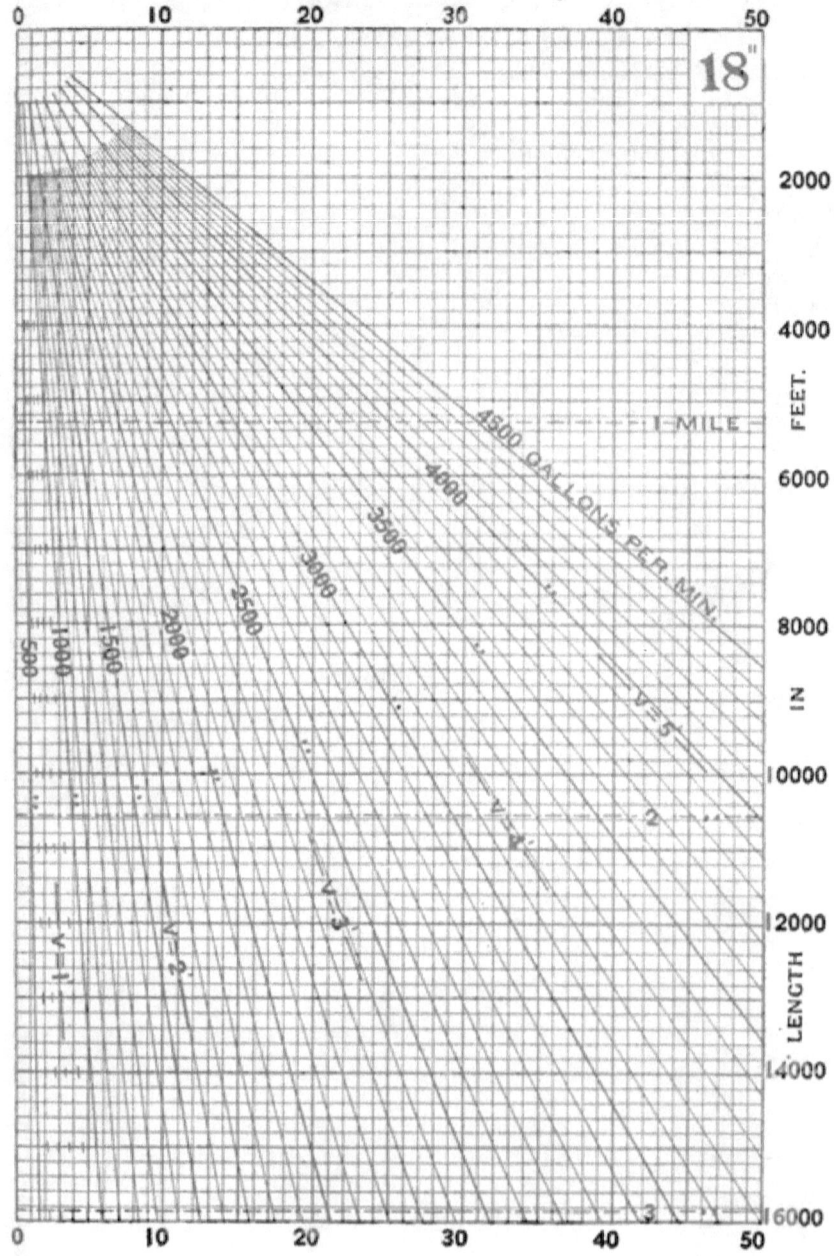

FLOW OF WATER IN LONG PIPES.

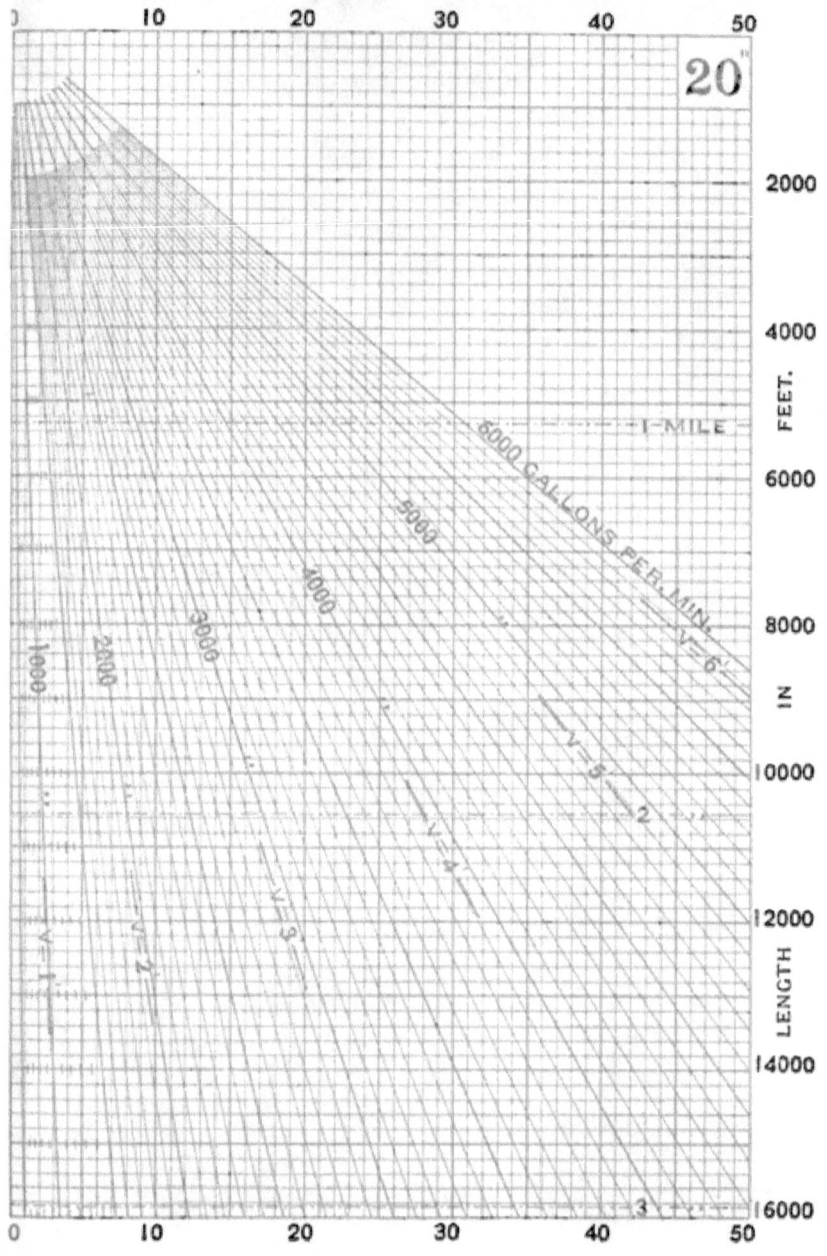

FLOW OF WATER IN LONG PIPES.

DIAGRAM No. 10

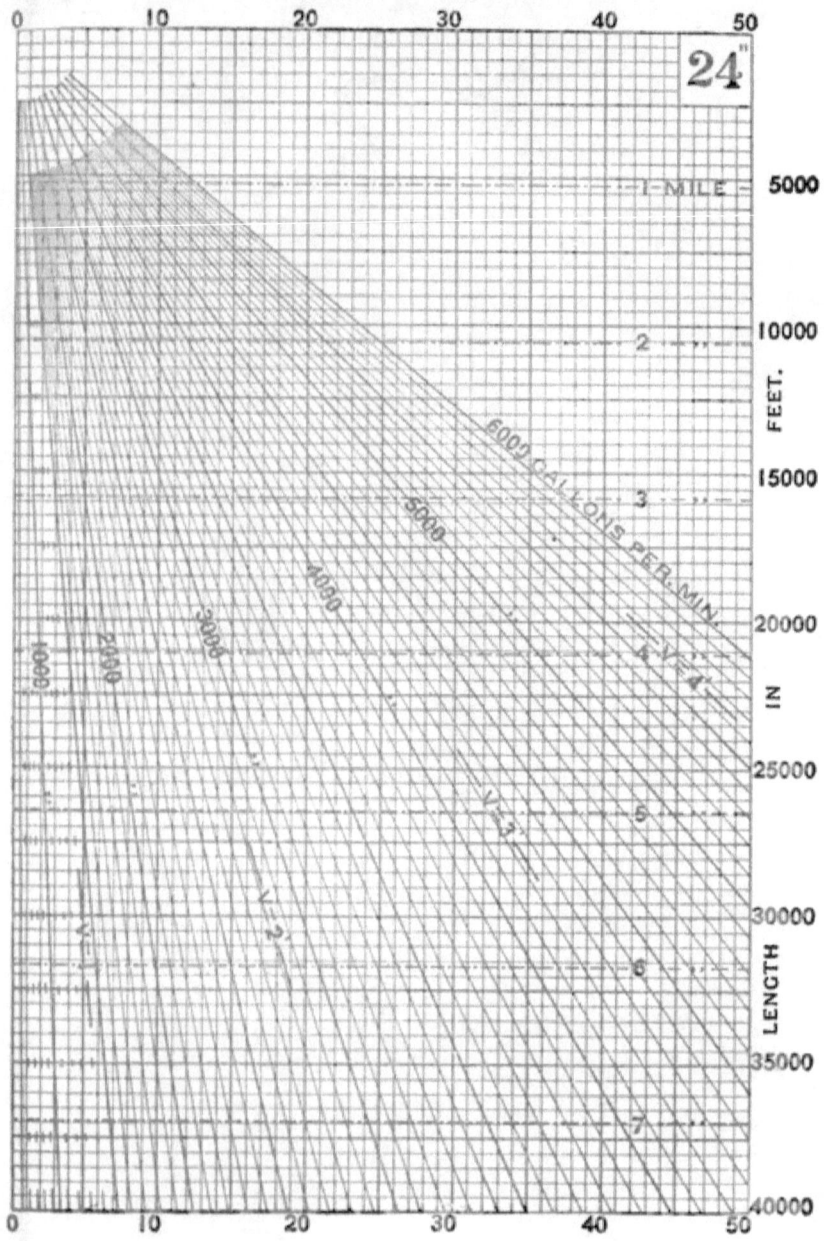

FLOW OF WATER IN LONG PIPES.

DIAGRAM No. 11

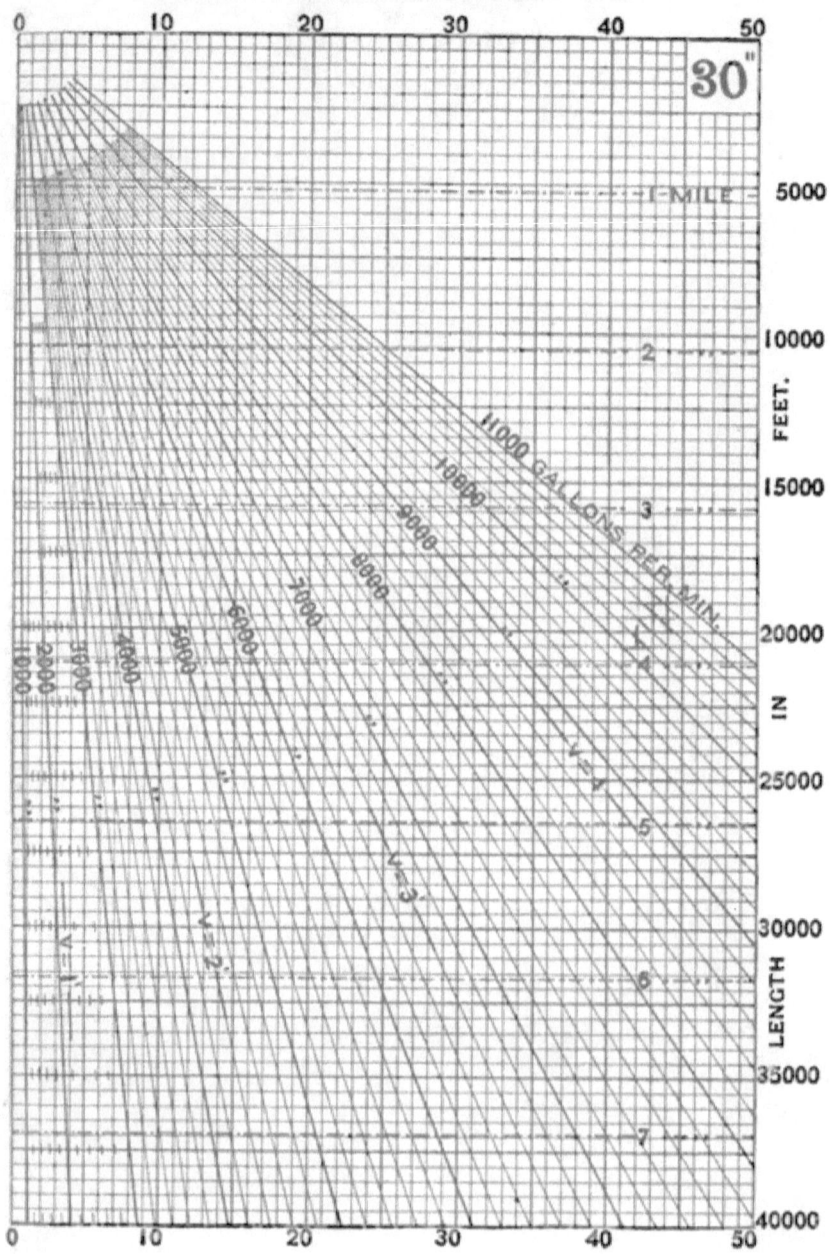

FLOW OF WATER IN LONG PIPES.

DIAGRAM No. 12

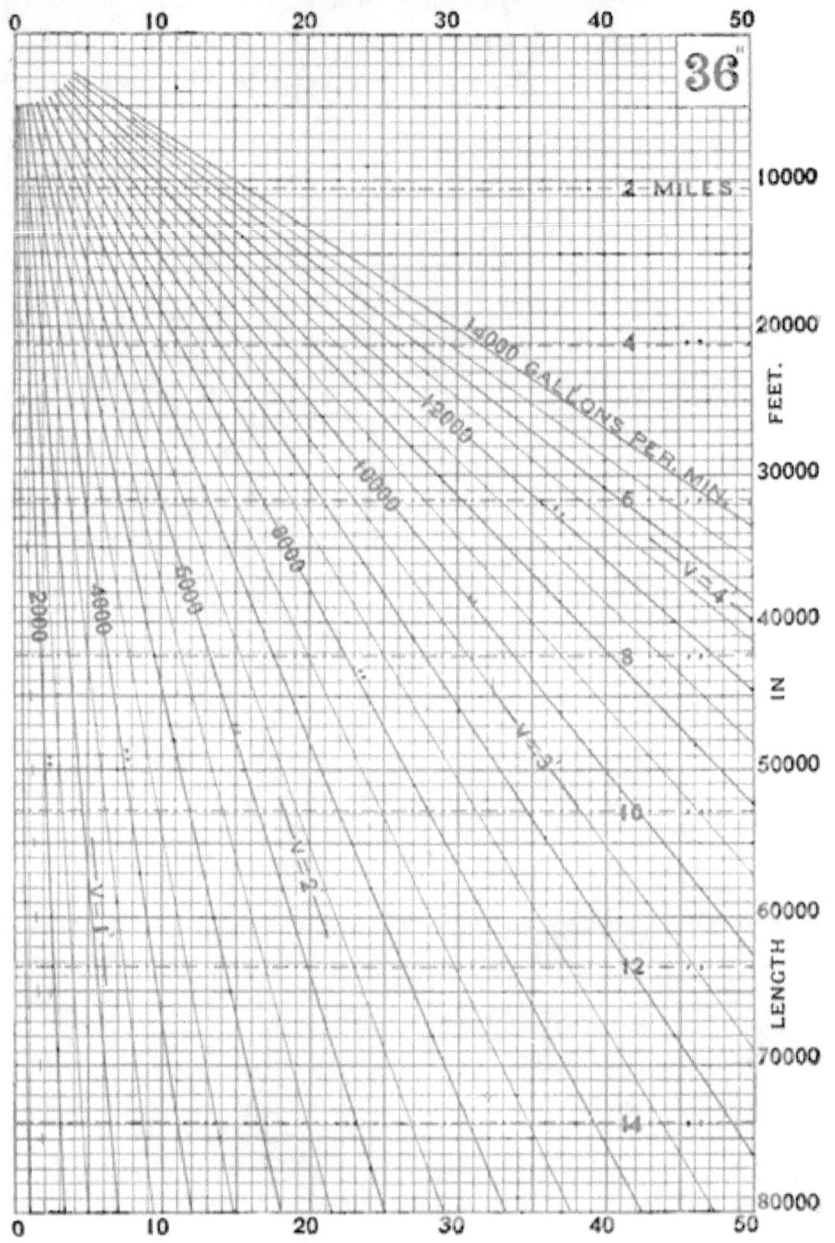

FLOW OF WATER IN LONG PIPES.

DIAGRAM No. 13

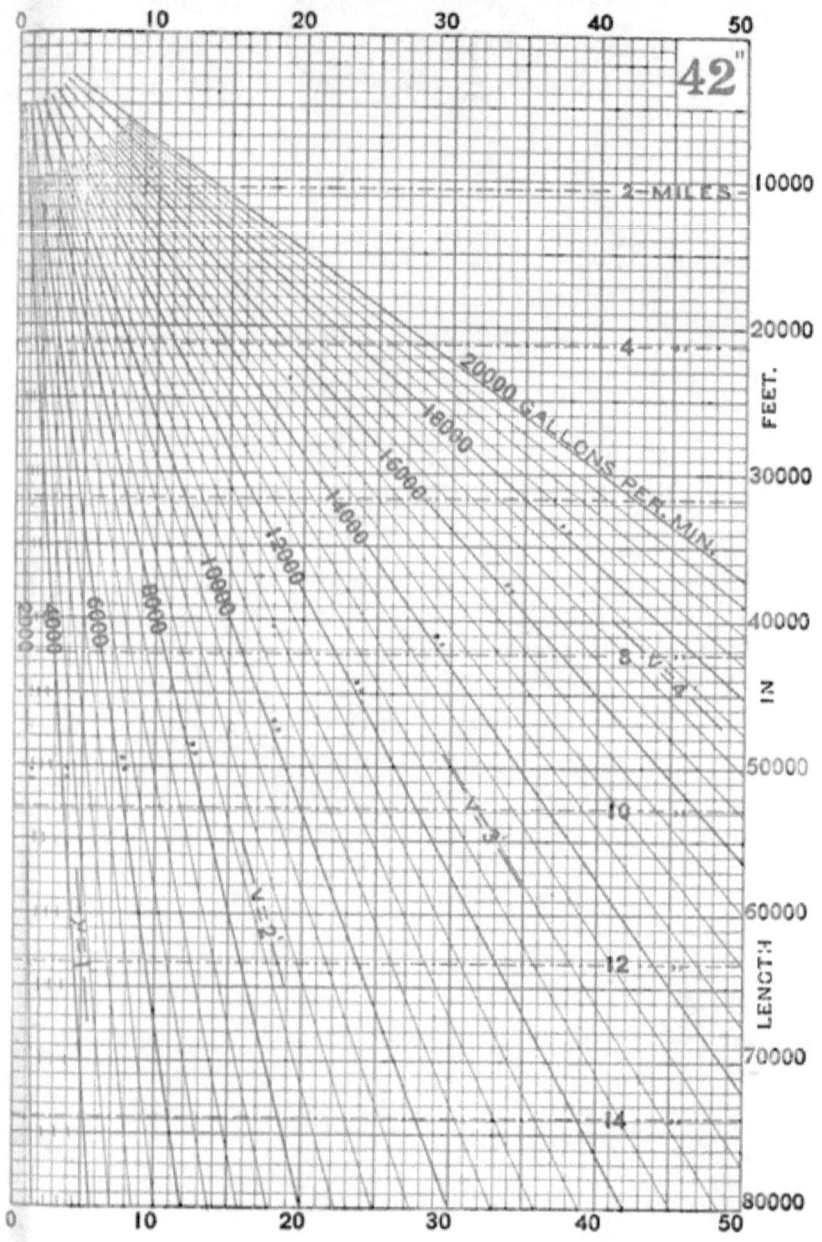

FLOW OF WATER IN LONG PIPES.

DIAGRAM No. 14

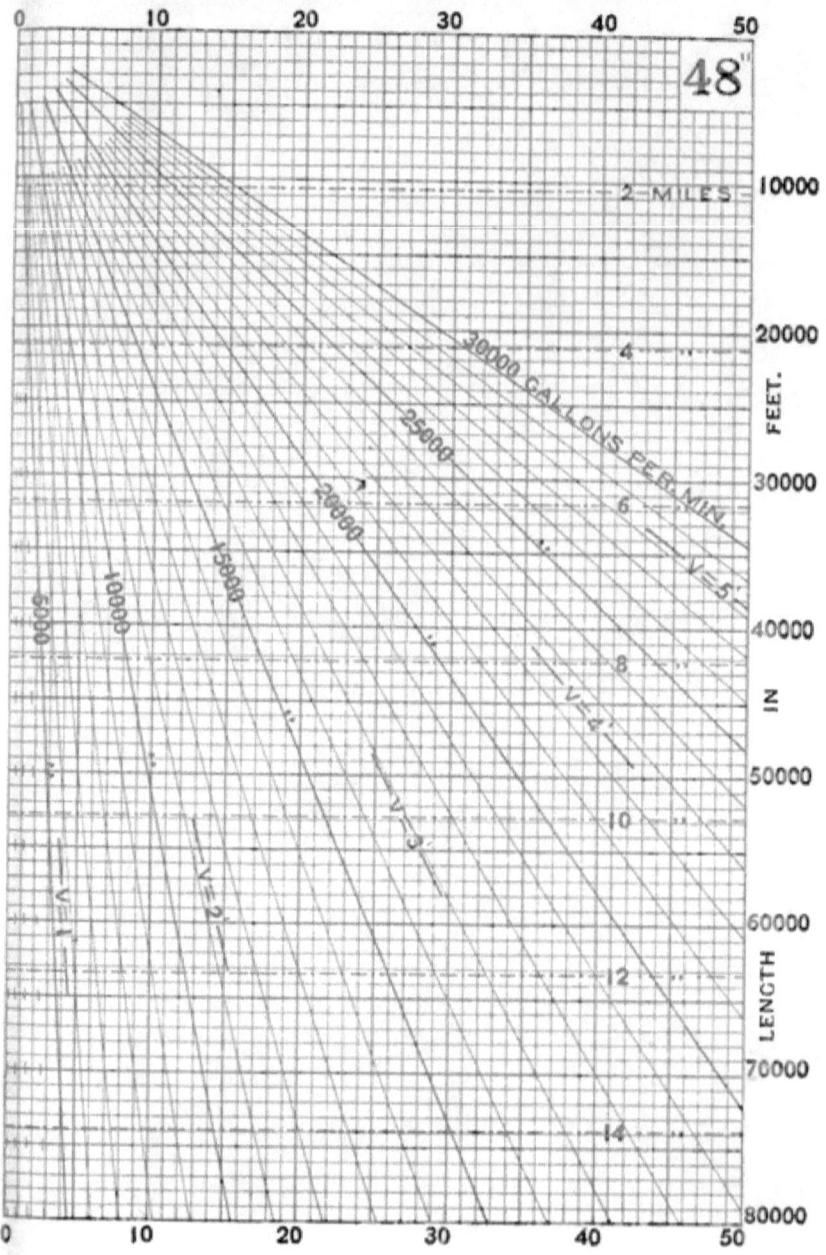

FLOW OF WATER IN LONG PIPES.

DIAGRAM No. 15

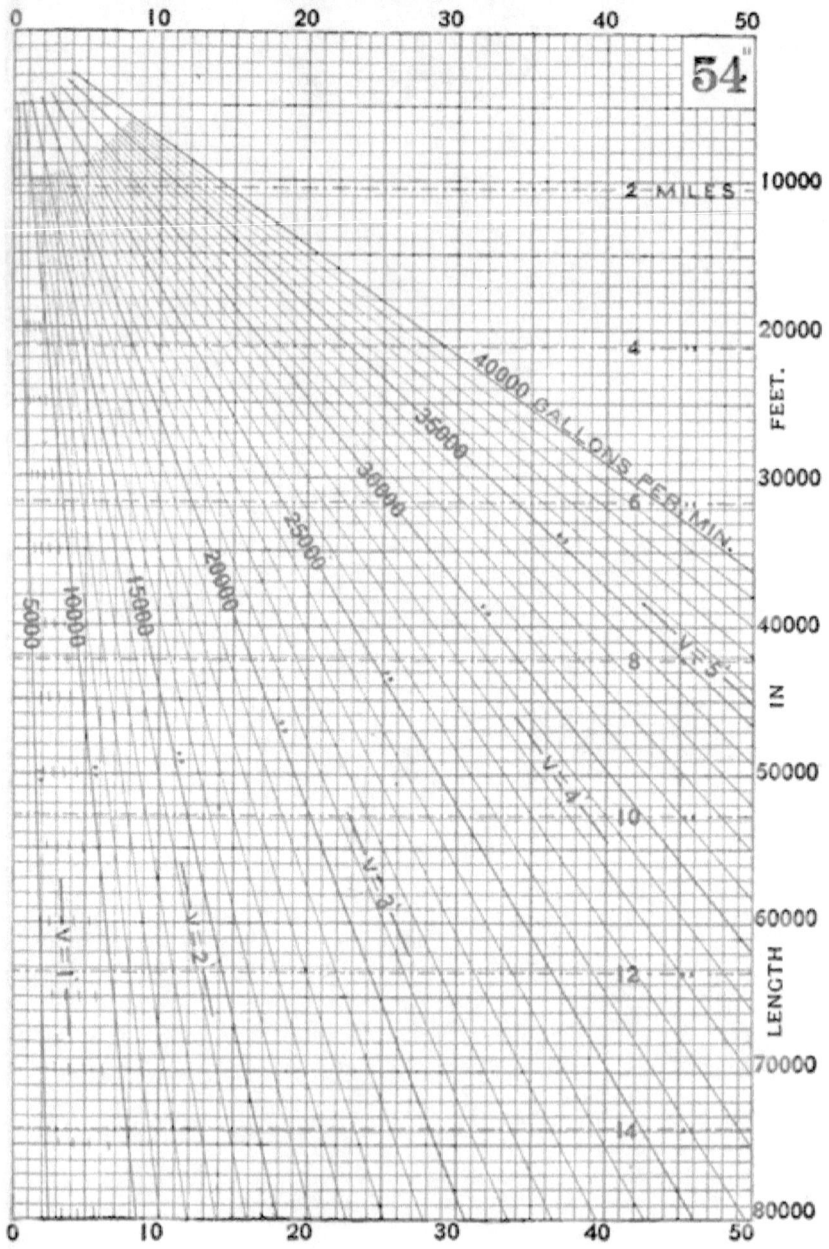

FLOW OF WATER IN LONG PIPES.

DIAGRAM No. 16

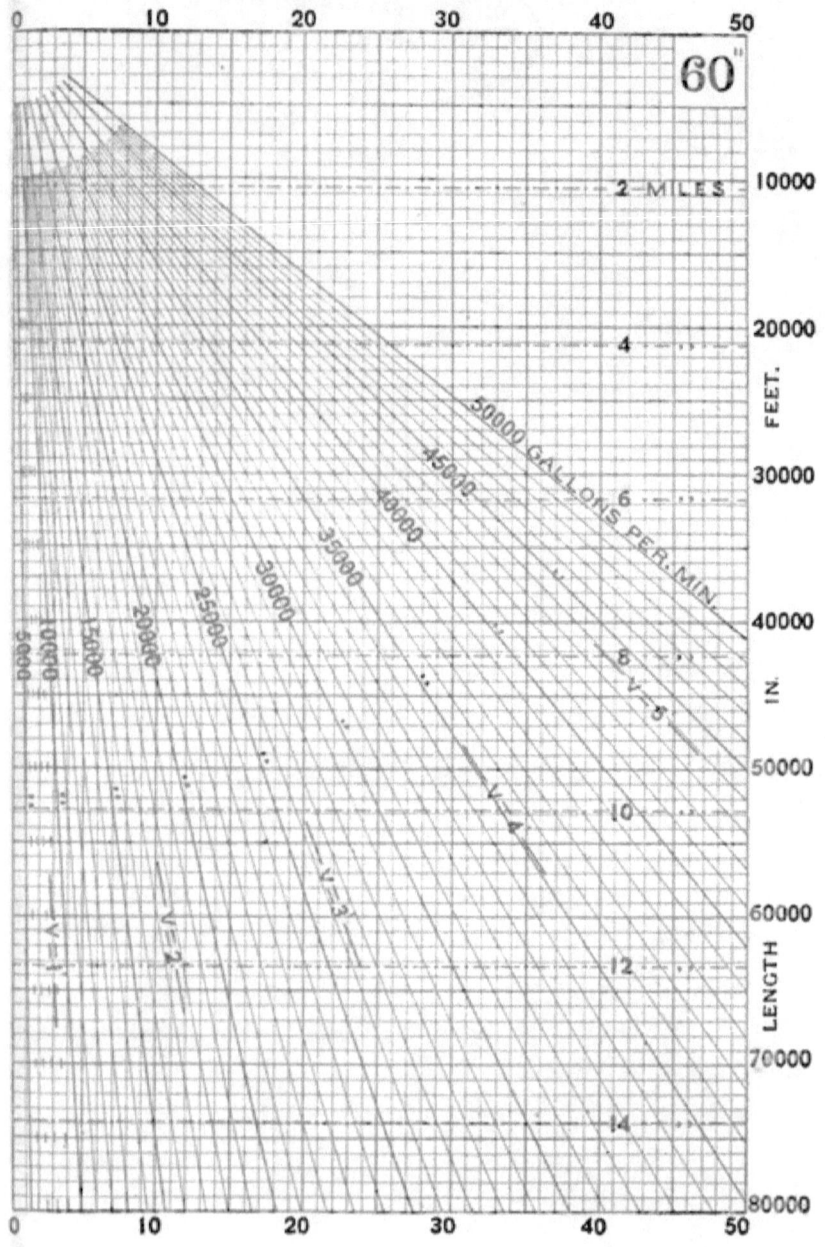

FLOW OF WATER IN LONG PIPES.

DIAGRAM No. 17

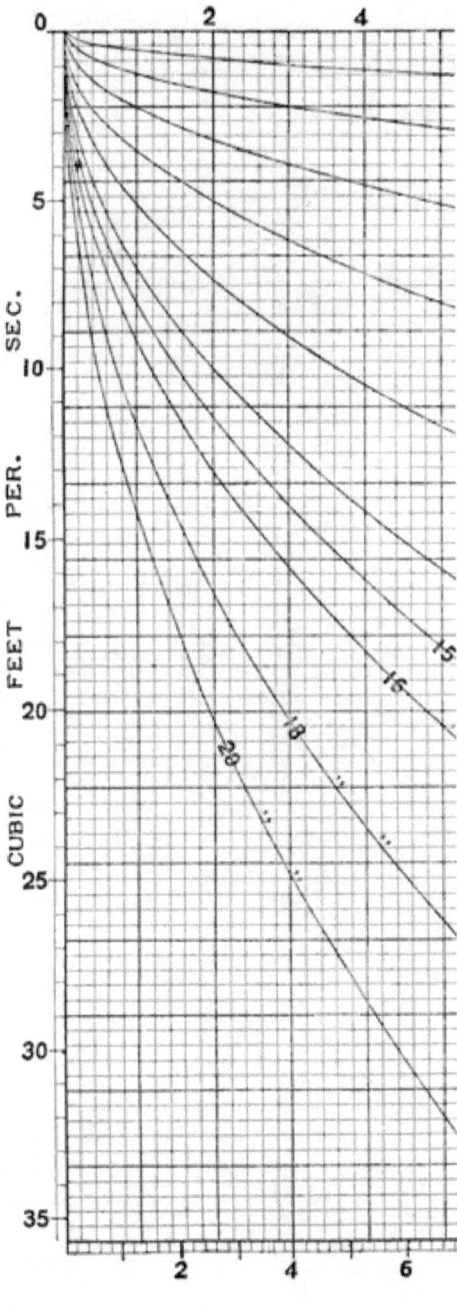

FLOW OF WAT

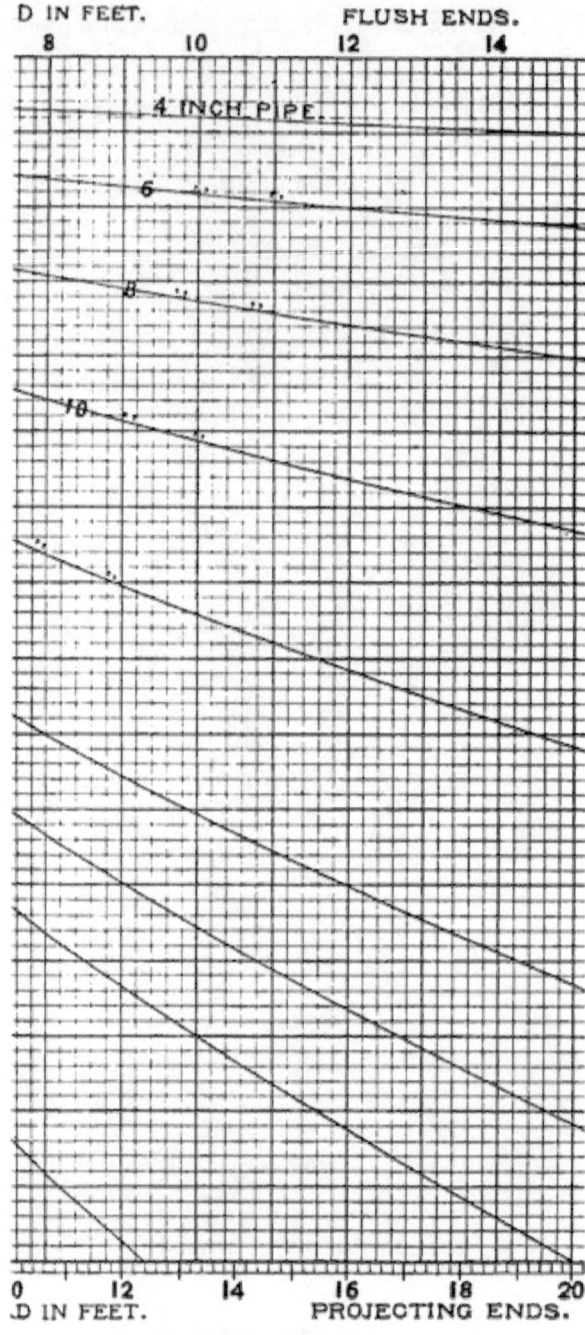

SHORT TUBES OR ENTRY HEAD.
DIAGRAM No. 18

2   4   6   8   10
                   HEAD

FLOW OF WATER IN SHO

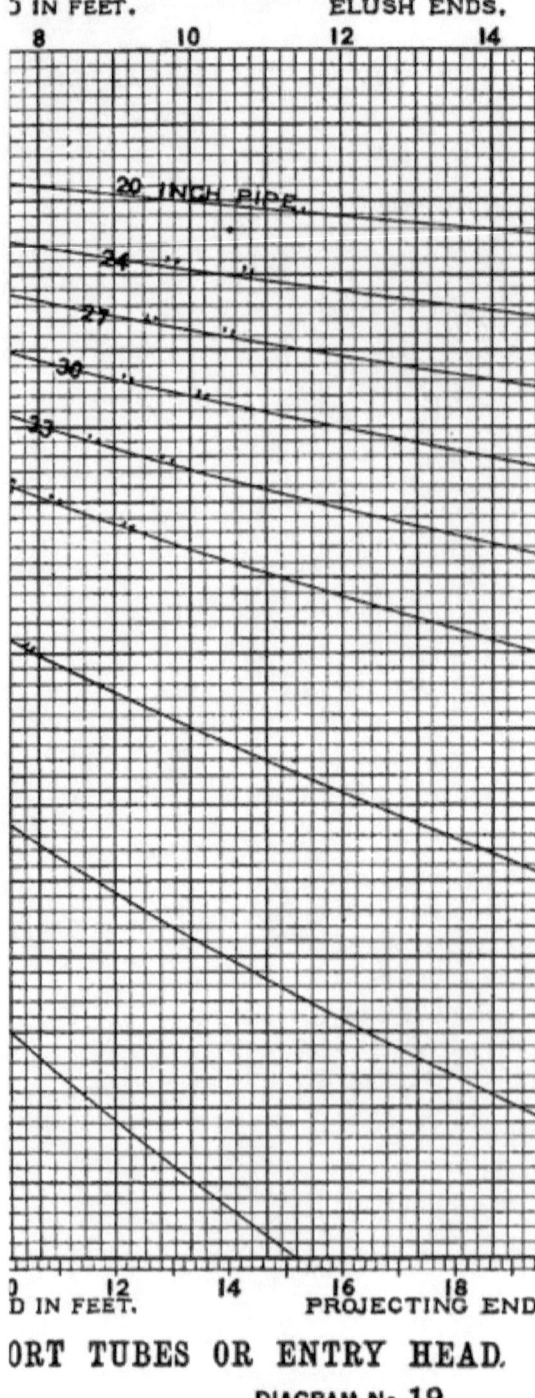

SHORT TUBES OR ENTRY HEAD.
DIAGRAM No. 19

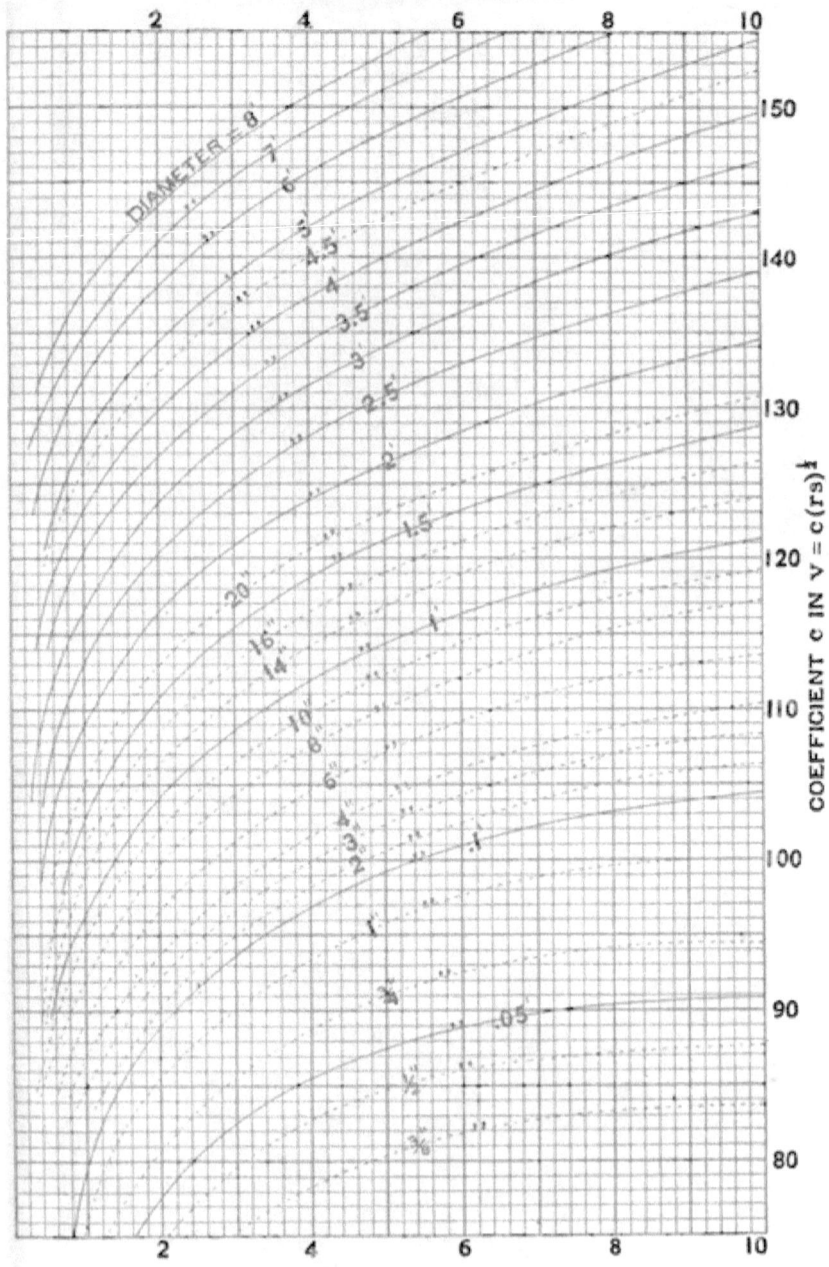

FROM SMITHS HYDRAULICS.   DIAGRAM No. 20

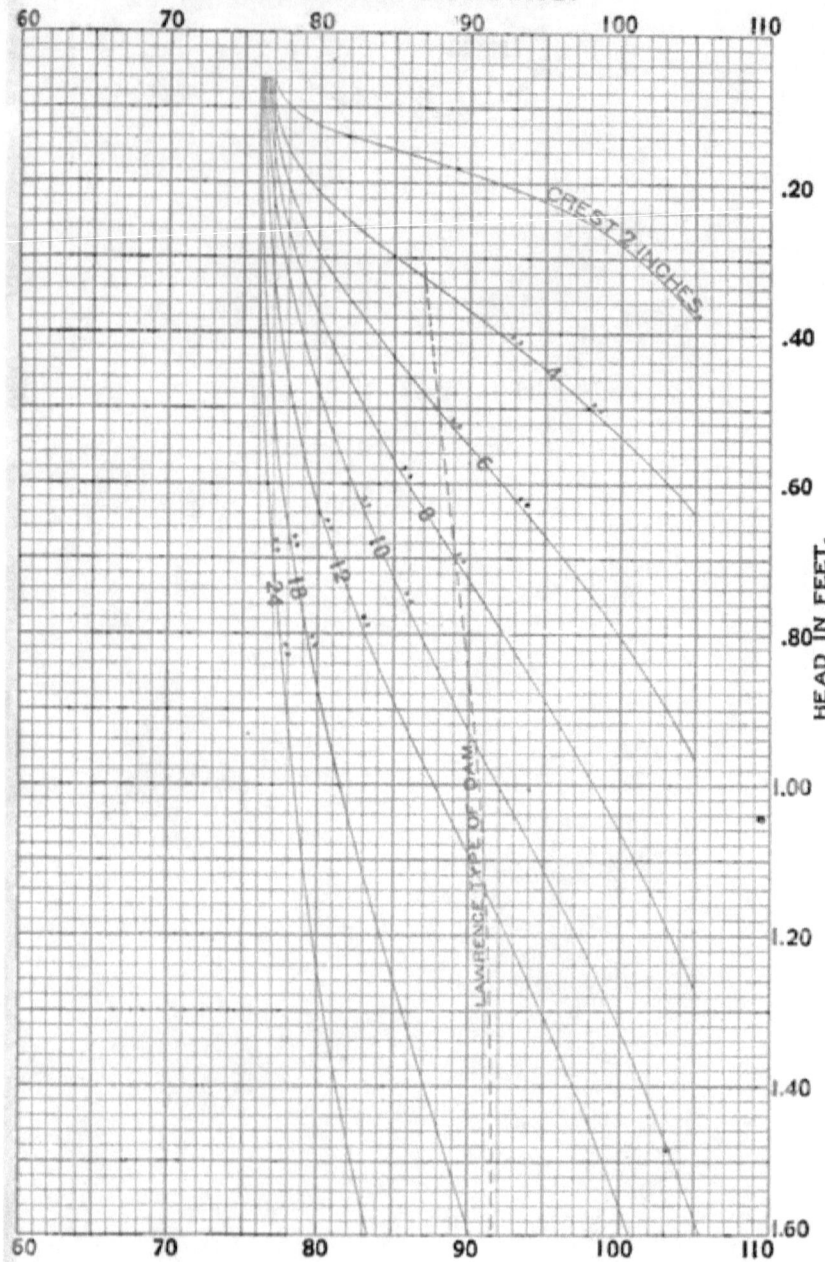

DISCHARGE OF WIDE CREST WEIRS.

DIAGRAM No. 21

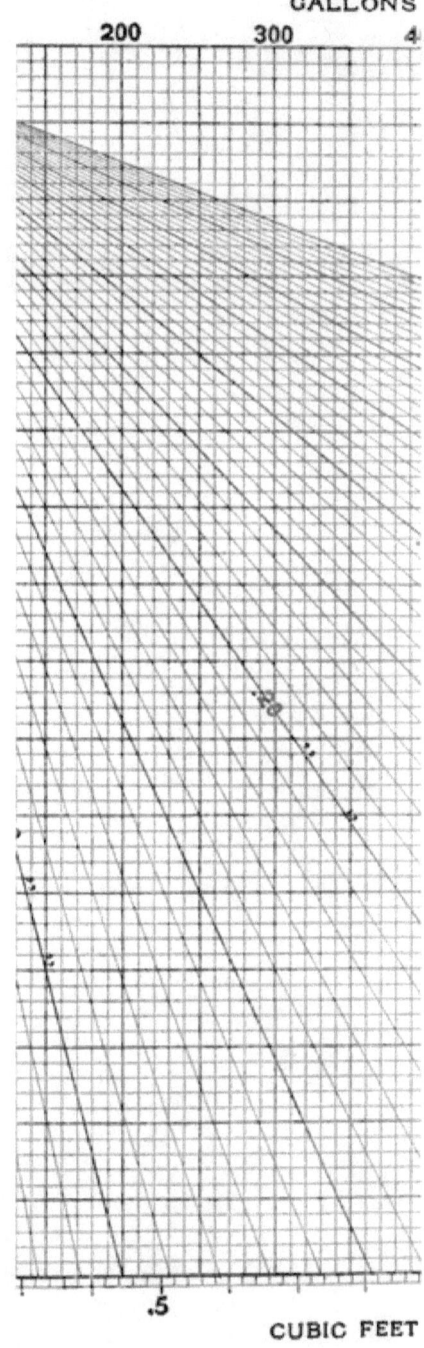

DISCHARGE OF RE(

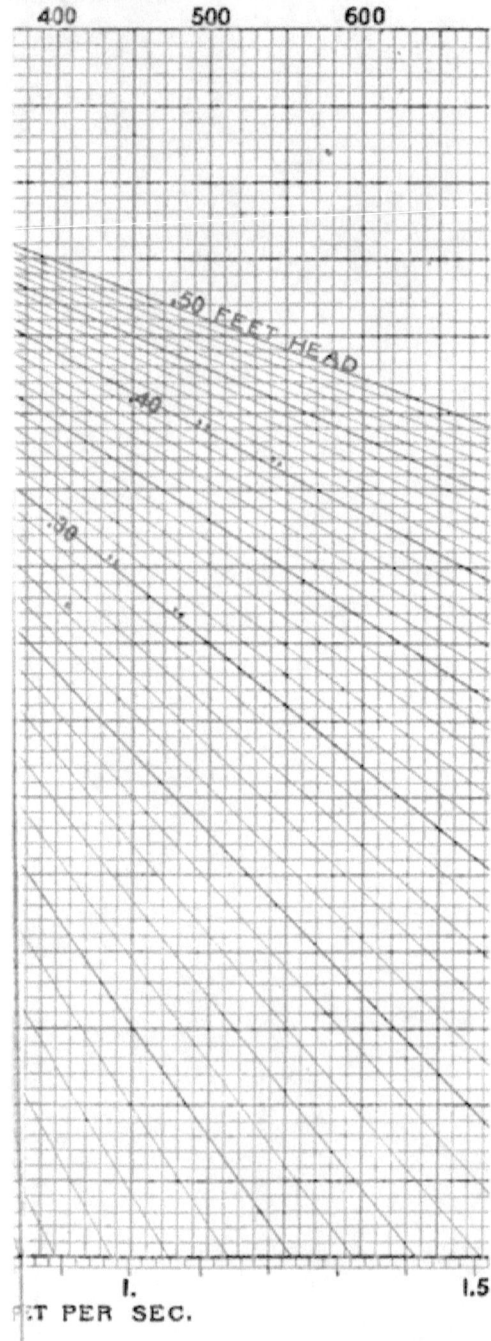

RECTANGULAR WEIRS.
DIAGRAM No. 22

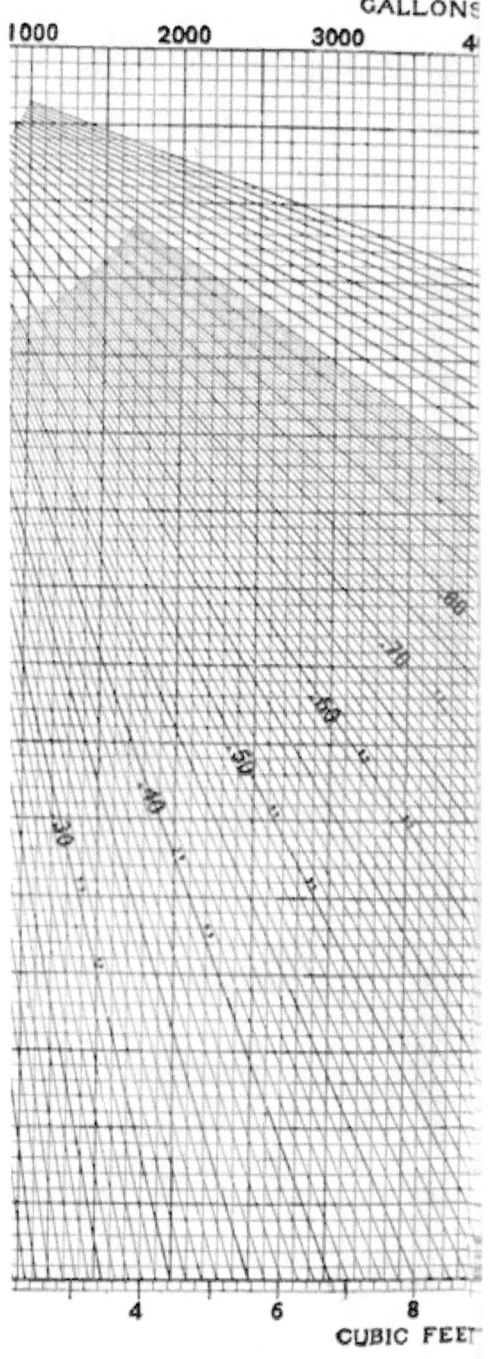

DISCHARGE OF REC

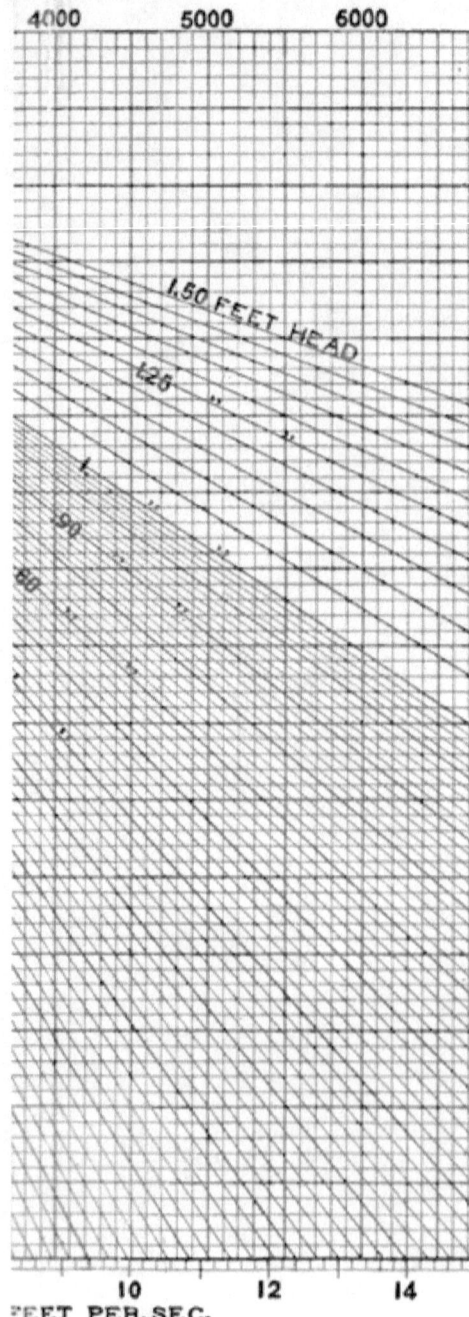

RECTANGULAR WEIRS.
DIAGRAM No. 23

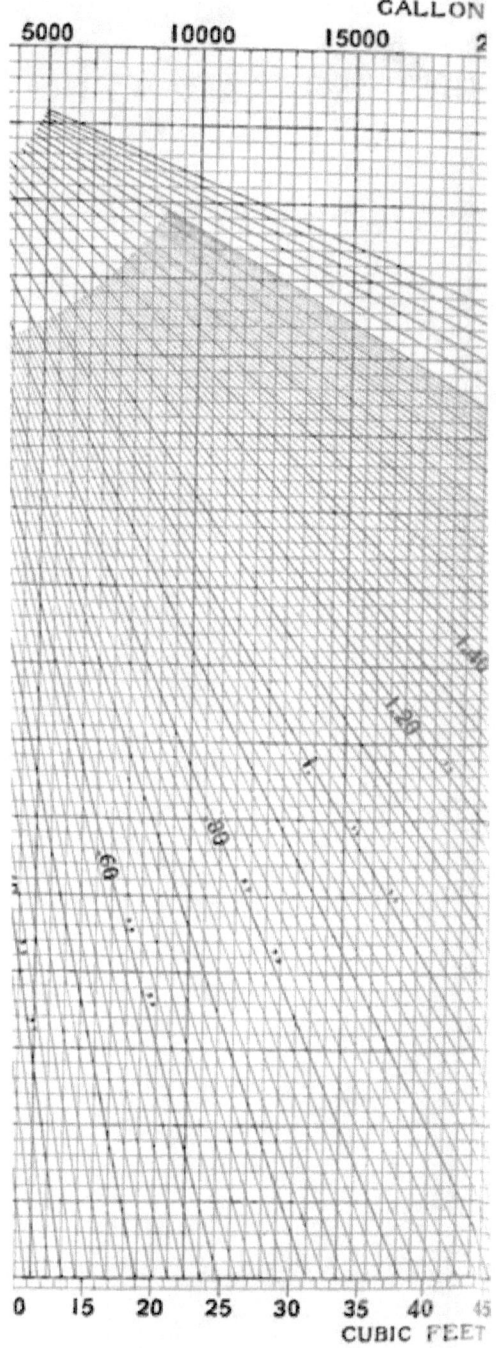

DISCHARGE OF REC

45 50 55 60 65 70 75
FEET PER. SEC.

RECTANGULAR WEIRS.
DIAGRAM No. 24

VELOCIT

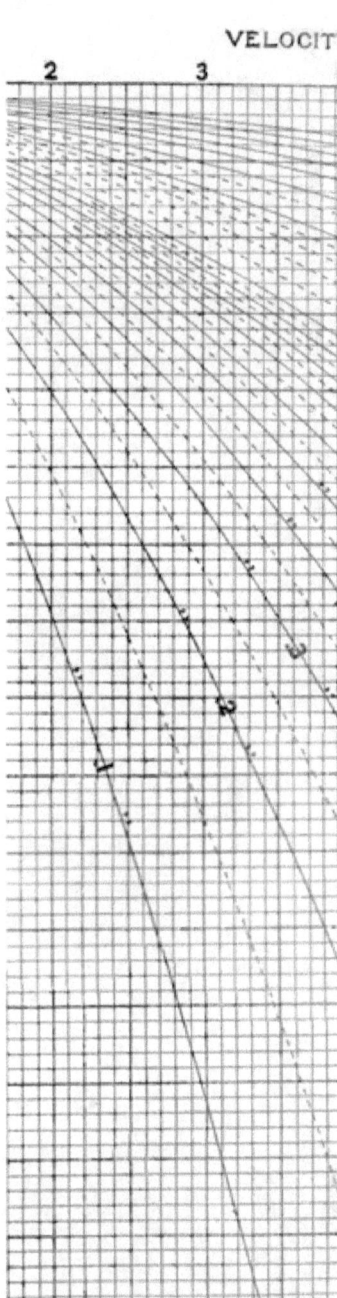

2   3
VELOCI

FLOW OF WATE

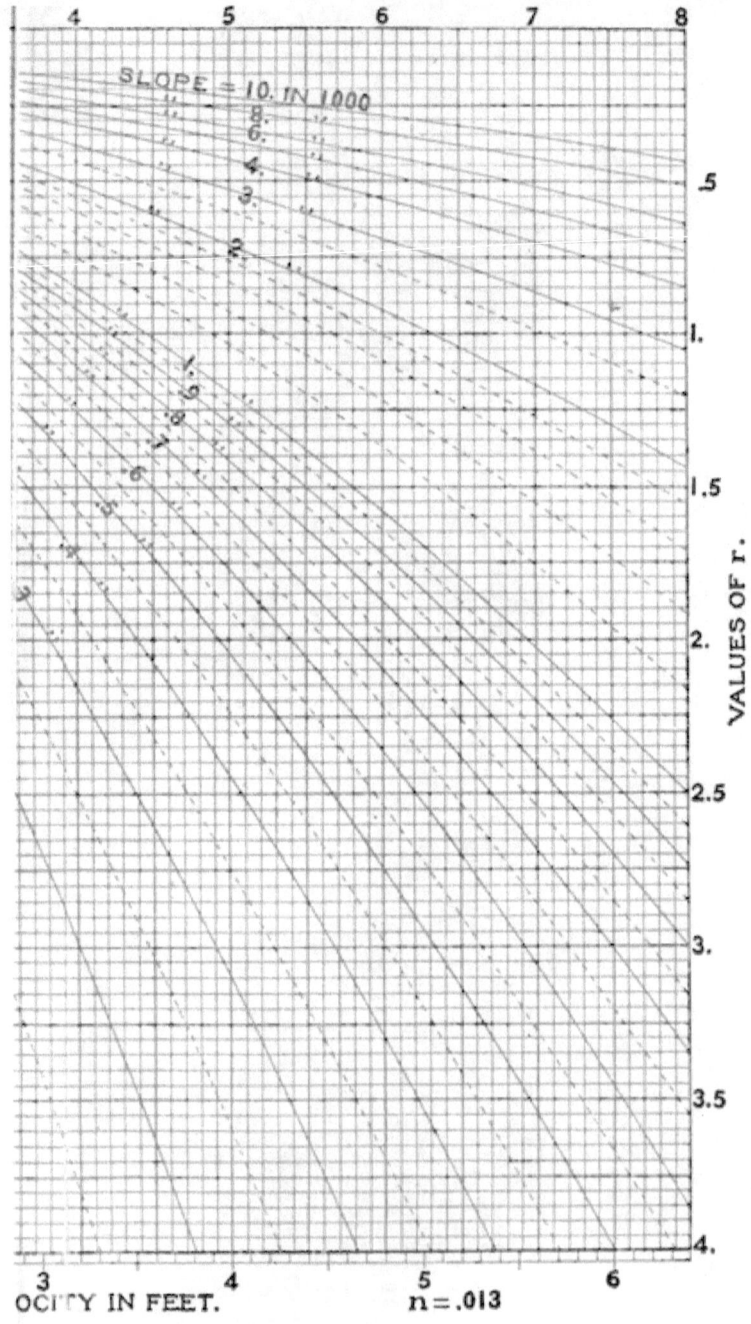

DIAGRAM No. 25

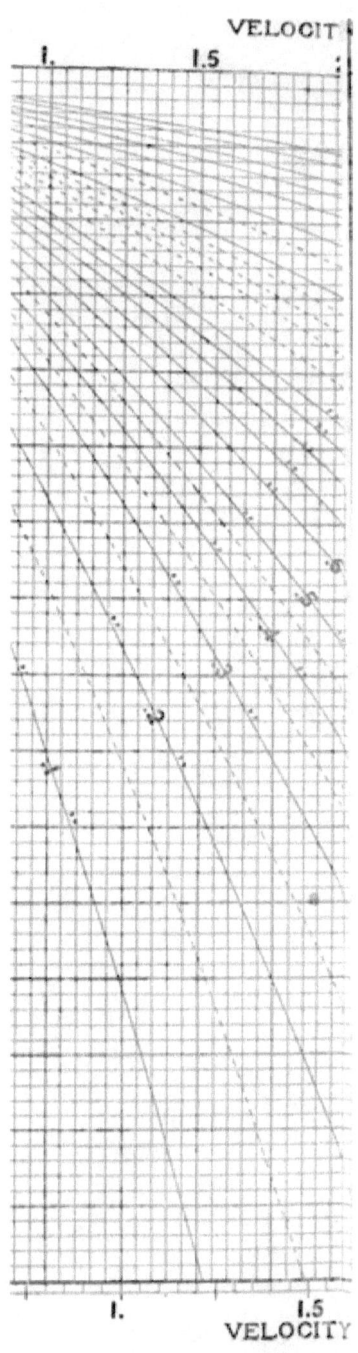

FLOW OF WATER

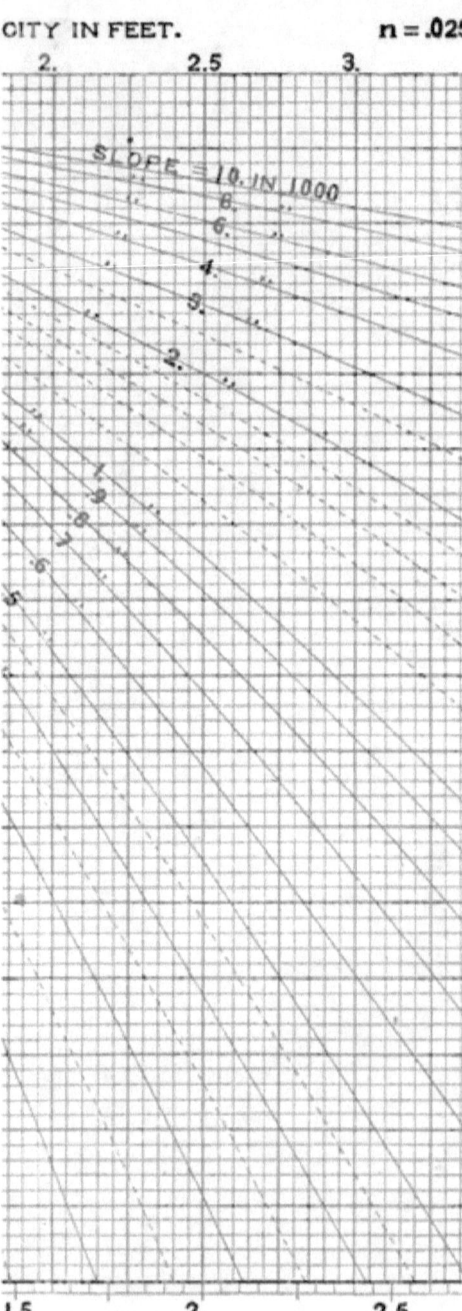

TER IN CHANNELS
DIAGRAM No. 26

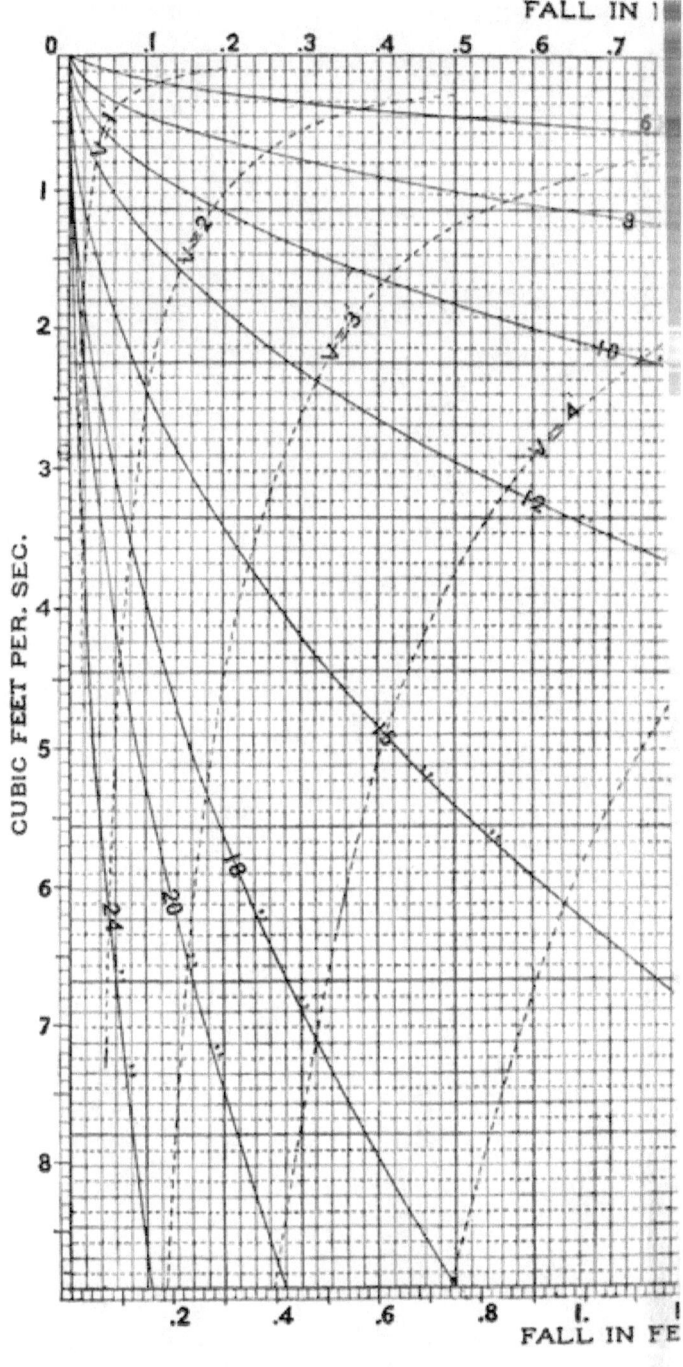

VELOCITY AND DISCHA

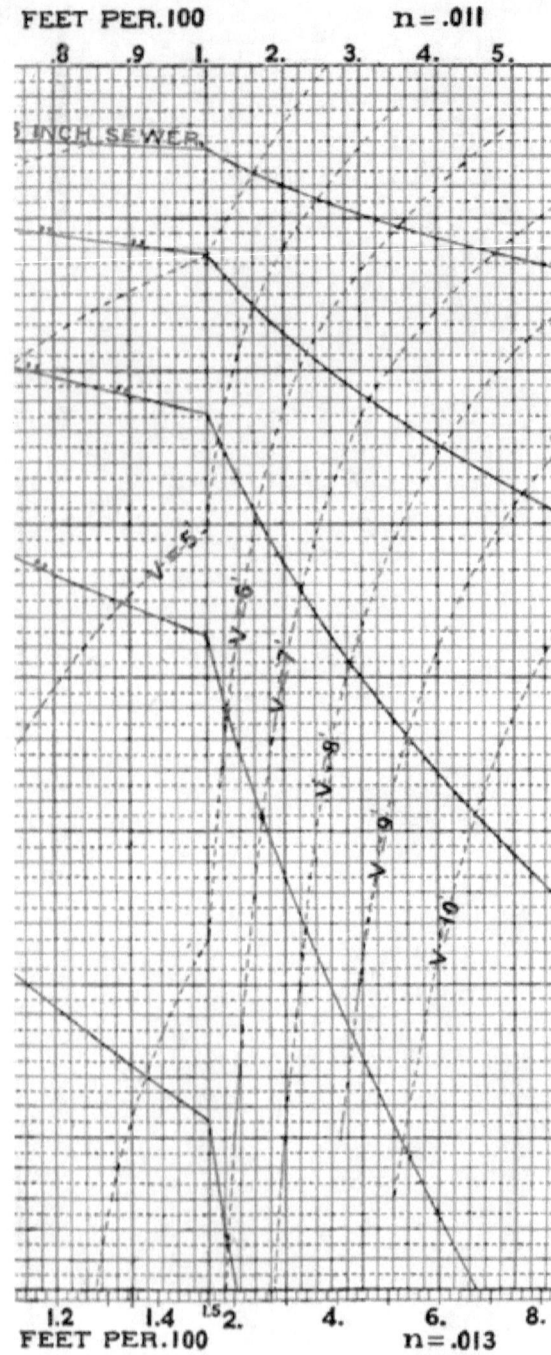

DISCHARGE OF PIPE SEWERS.
DIAGRAM No. 27

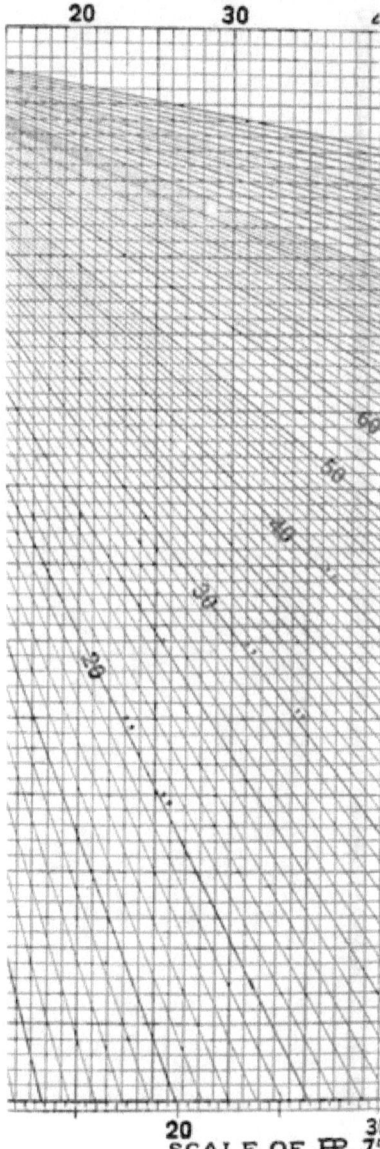

HORSE

GIVING HORSE POWER OF FALLING
REQUIRED TO PUMP WATER TO DI.
WEIGHT OF WATER TAKEN AT

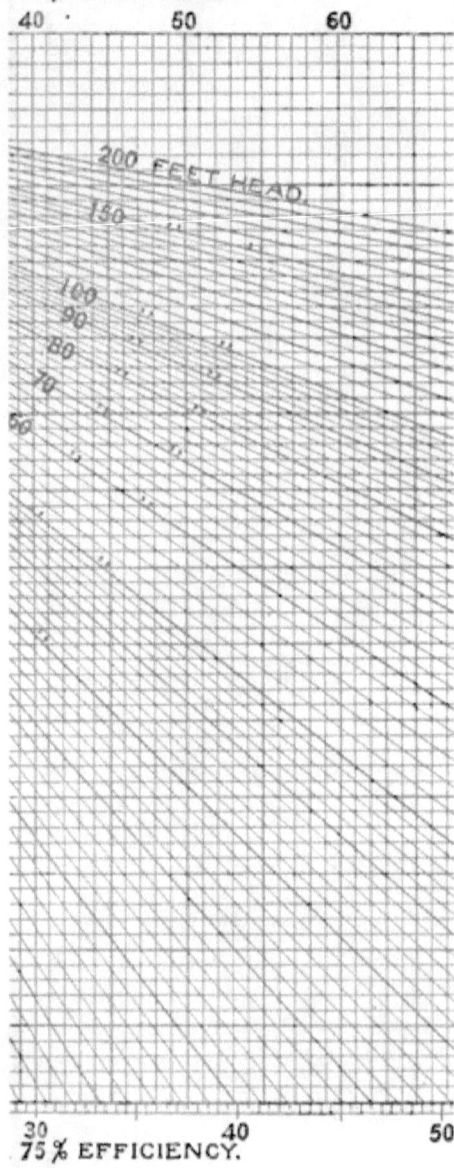

POWER

ING WATER ALSO OF HORSE POWER
DIFFERENT HEIGHTS

62.4 LBS. PER. CUBIC FOOT.

DIAGRAM No. 28

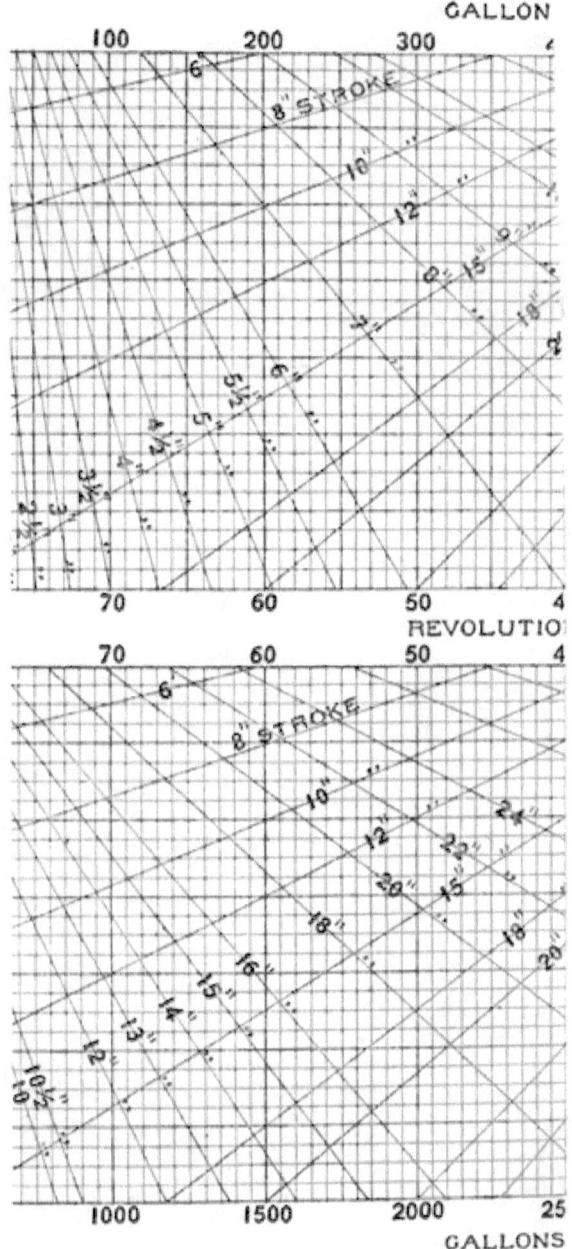

DISCHARGE OF

FOR DUPLEX MULTIPLY BY 2.

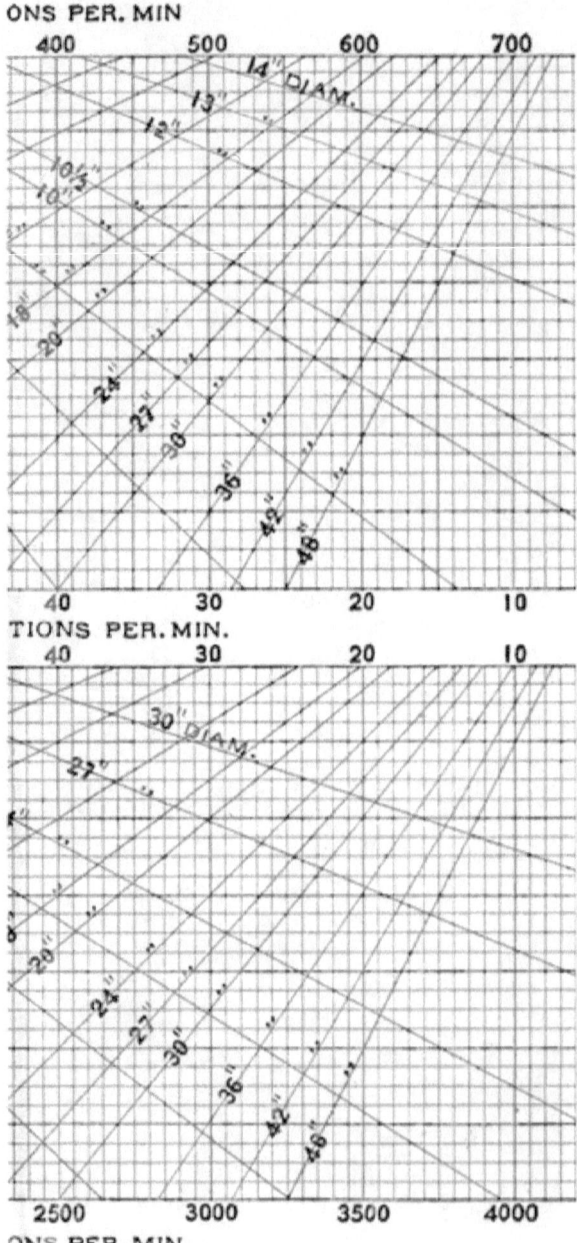

OF SINGLE PUMPS.

DIAGRAM No. 29

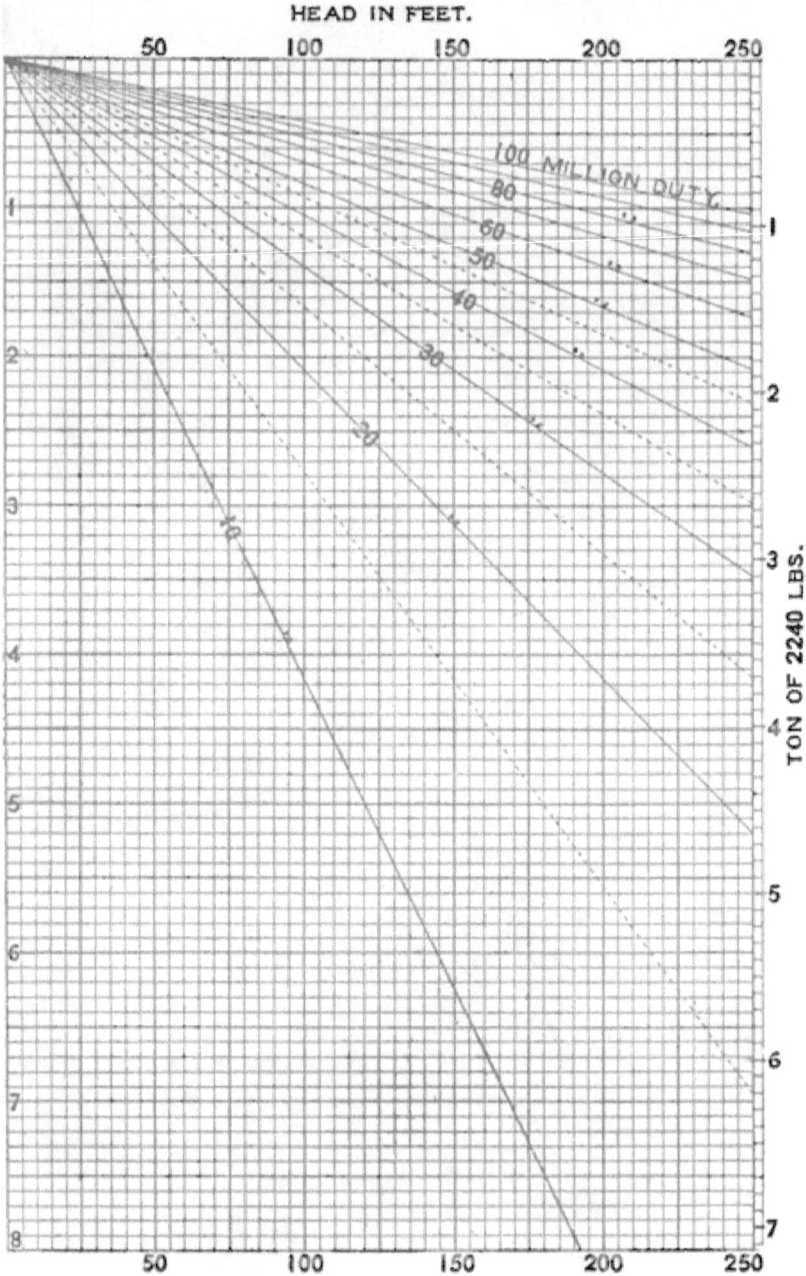

COAL REQUIRED IN PUMPING.
1000000 GALLONS OF WATER.   DIAGRAM No. 30

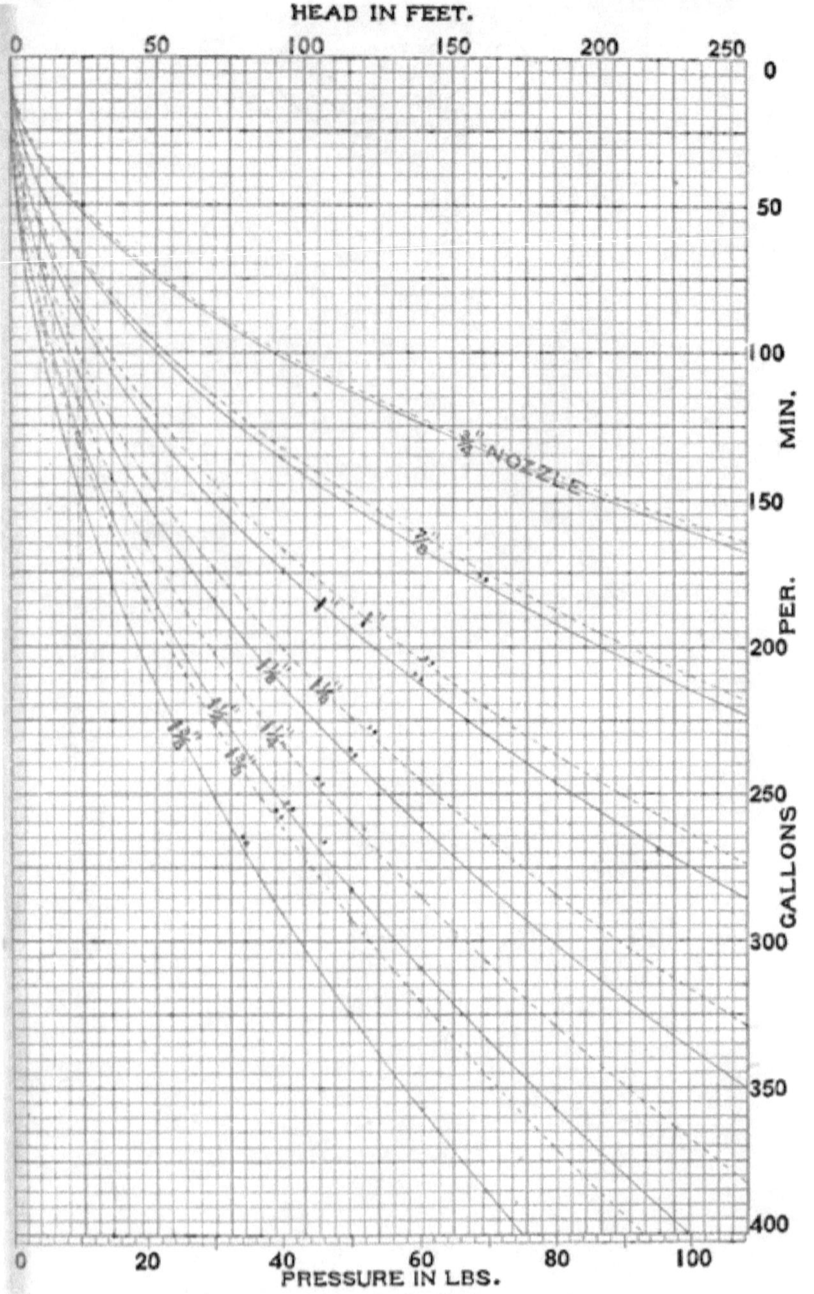

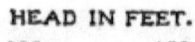

DISCHARGE OF NOZZLES.

FULL CURVE FOR 50' HOSE.   PRESSURE INDICATED AT HYDRANT.
DOTTED "   " 100' "   DIAGRAM No. 31

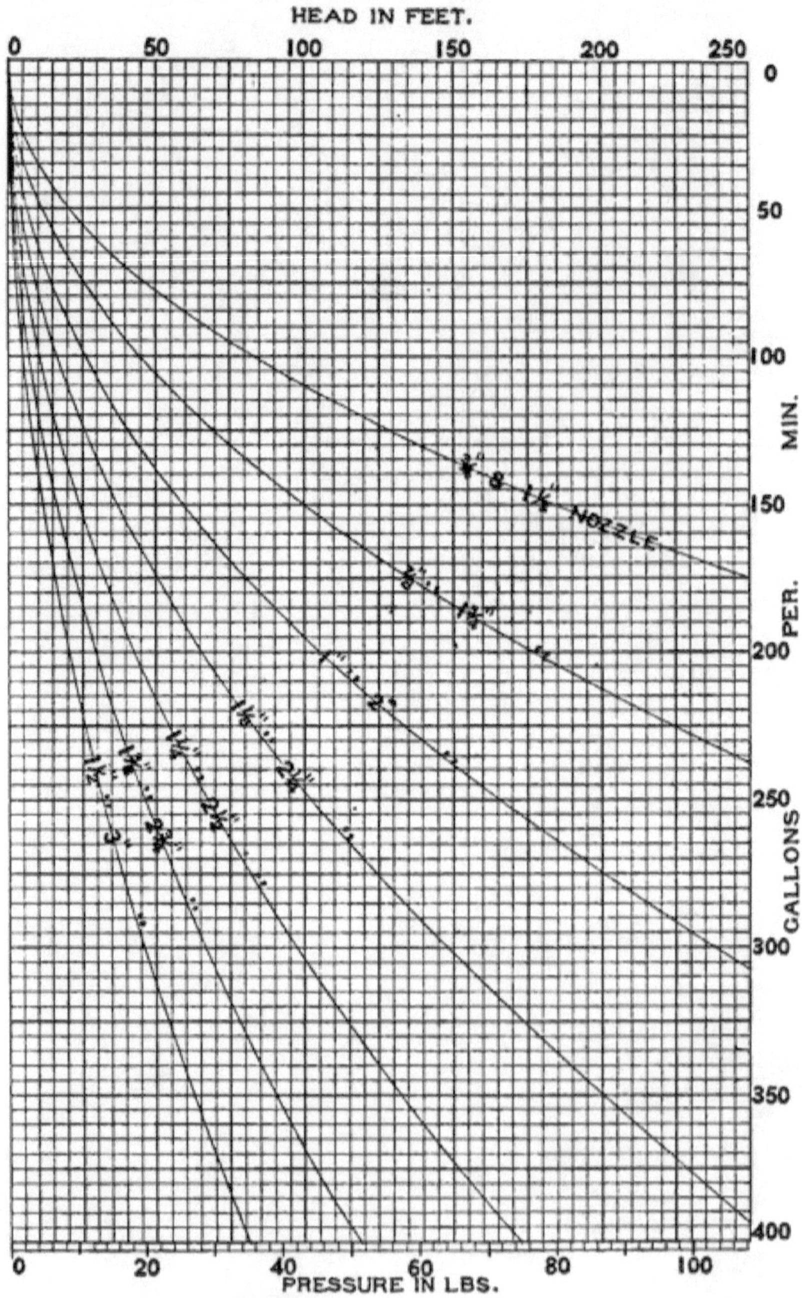

DISCHARGE OF NOZZLES.
PRESSURE INDICATED AT BASE OF PLAY PIPE.
FOR LARGE NOZZLES MULTIPLY GALLONS BY 4.

DIAGRAM No. 32

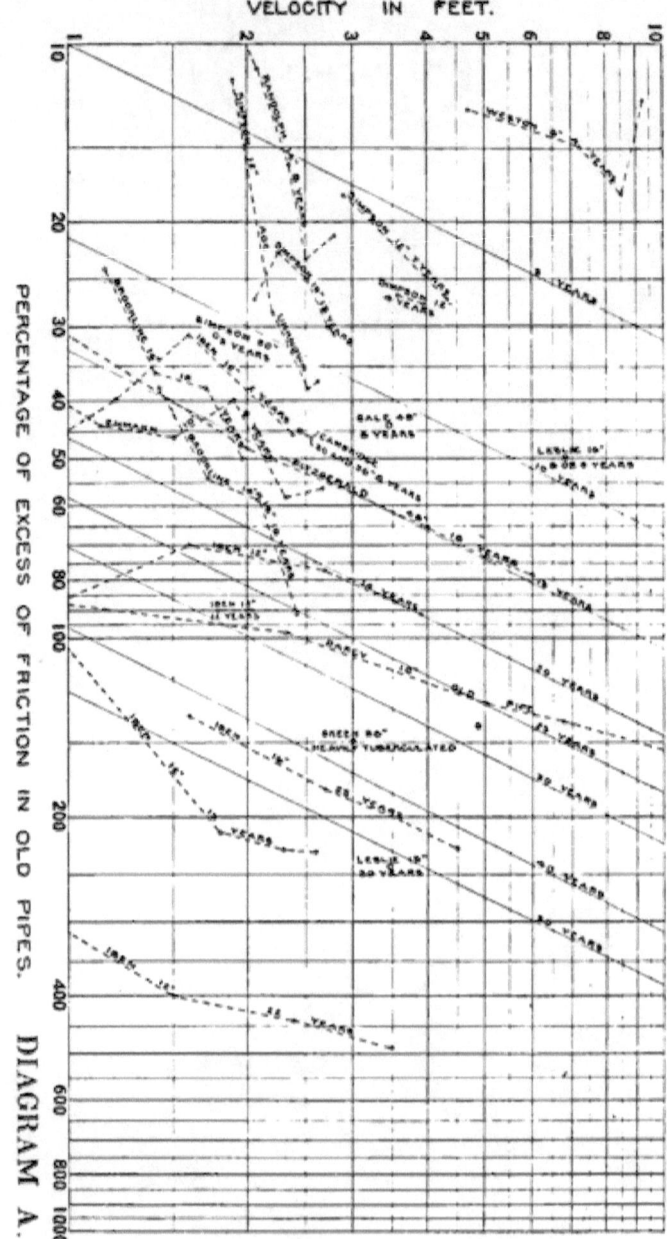

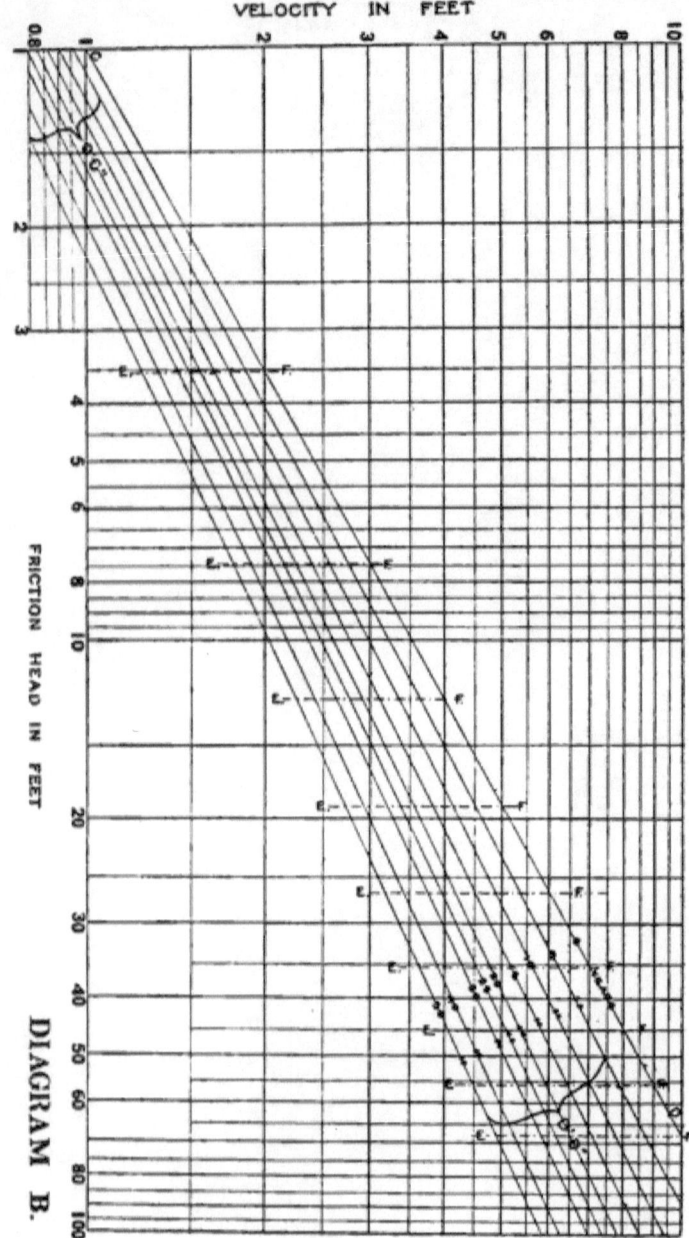

DIAGRAM B.

www.ingramcontent.com/pod-product-compliance
Lightning Source LLC
Chambersburg PA
CBHW020313170426
43202CB00008B/585